Mission Command:
Command and Control of Army Forces

July 2019

United States Government
US Army

Contents

DISTRIBUTION RESTRICTION: Approved for public release; distribution is unlimited.

This publication supersedes ADP 6-0, dated 17 May 2012, and ADRP 6-0, dated 17 May 2012.

Figures

Tables

Vignettes

Preface

ADP 6-0, *Mission Command: Command and Control of Army Forces*, provides a discussion of the fundamentals of mission command, command and control, and the command and control warfighting function. It describes how commanders, supported by their staffs, combine the art and science of command and control to understand situations, make decisions, direct actions, and lead forces toward mission accomplishment.

To comprehend the doctrine contained in ADP 6-0, readers should understand the nature of operations and the fundamentals of unified land operations described in ADP 3-0, *Operations*. Army leadership attributes and competencies are vital to exercising command and control, and readers should also be familiar with the fundamentals of leadership in ADP 6-22, *Army Leadership*, and FM 6-22, *Leader Development*. The Army Ethic guides decisions and actions while exercising command and control, and readers must understand the ideas in ADP 1-1, *The Army Profession*. As the operations process is the framework for exercising command and control, readers must also understand the fundamentals of the operations process established in ADP 5-0, *The Operations Process*.

The doctrine in ADP 6-0 forms the foundation for command and control tactics, techniques, and procedures. For an explanation of these tactics and procedures, see FM 6-0, *Commander and Staff Organization and Operations*. For an explanation of the techniques associated with command and control, see ATP 6-0.5, *Command Post Organization and Operations,* as well as other supporting techniques publications.

The principal audience for ADP 6-0 is Army commanders, leaders, and unit staffs. Mission command demands more from subordinates at all levels, and understanding and practicing the mission command principles during operations and garrison activities are imperative for all members of the Army Profession.

The Army historically fights with joint and multinational partners as part of a coalition, and ADP 6-0 is nested with joint and multinational doctrine. Commanders and staffs of Army headquarters that require joint capabilities to conduct operations, or serving as a joint task force or multinational headquarters, should also refer to applicable doctrine concerning command and control of joint or multinational forces.

ADP 6-0 implements North Atlantic Treaty Organization standardization agreement 2199, *Command and Control of Allied Land Forces.*

Commanders, staffs, and subordinates ensure their decisions and actions comply with applicable U.S., international, and, in some cases, host-nation laws and regulations. Commanders at all levels ensure their Soldiers operate in accordance with the Army Ethic, the law of war, and the rules of engagement. (See FM 27-10 for a discussion of the law of war.)

ADP 6-0 uses joint terms where applicable. Selected joint and Army terms and definitions appear in both the glossary and the text. Terms for which ADP 6-0 is the proponent publication (the authority) are marked with an asterisk (*) in the glossary. Definitions for which ADP 6-0 is the proponent publication are boldfaced in the text. For other definitions shown in the text, the term is italicized and the number of the proponent publication follows the definition.

ADP 6-0 applies to the Active Army, Army National Guard/Army National Guard of the United States, and United States Army Reserve unless otherwise stated.

The proponent of ADP 6-0 is the United States Army Combined Arms Center. The preparing agency is the Combined Arms Doctrine Directorate, Mission Command Center of Excellence. Send comments and recommendations on a DA Form 2028 (*Recommended Changes to Publications and Blank Forms*) to Commander, U.S. Army Combined Arms Center and Fort Leavenworth, ATTN: ATZL-MCD (ADP 6-0), 300 McPherson Avenue, Fort Leavenworth, KS 66027-2337; by email to usarmy.leavenworth.mccoe.mbx.cadd-org-mailbox@mail.mil; or submit an electronic DA Form 2028.

This page intentionally left blank.

Acknowledgements

The copyright owners listed here have granted permission to reproduce material from their works. The Source Notes lists other sources of quotations and photographs.

Excerpts from *On War* by Carl von Clausewitz. Edited and translated by Peter Paret and Michael E. Howard. Copyright © 1976, renewed 2004 by Princeton University Press.

Quotes reprinted courtesy B. H. Liddell Hart, *Strategy*, 2d rev. ed. Copyright © 1974 by Signet Printing. Copyright © renewed 1991 by Meridian.

Excerpts from *War as I Knew It* by General George S. Patton. Copyright © 1947 by Beatrice Patton Walters, Ruth Patton Totten, and George Smith Totten. Copyright © renewed 1975 by MG George Patton, Ruth Patton Totten, John K. Waters, Jr., and George P. Waters. Reprinted by permission of Houghton Mifflin Company. All rights reserved.

Quote reprinted courtesy Field-Marshall Viscount William Slim, *Defeat into Victory: Battling Japan in Burma and India, 1942–1945*. Copyright © 1956 by Viscount William Slim. Copyright © renewed 2000 by Copper Square Press.

Quote courtesy Logan Nye, "How the 'Little Groups of Paratroopers' Became Airborne Legends," *We Are the Mighty*, 8 April 2016. Online http://freerepublic.com/focus/f-chat/3535576/posts?page=12.

Quote courtesy Field-Marshal Earl Wavell, *Soldiers and Soldiering or Epithets of War*. Oxford, United Kingdom: Alden Press, 1953.

Excerpts from Matthew B. Ridgway, *Soldier: The Memoirs of Matthew B. Ridgway*. Copyright © 1956 by Matthew B. Ridgway. Copyright © 1956 The Curtis Publishing Company. Reprinted by permission of Andesite Press, 2017.

Quote courtesy Gary Klein, *Sources of Power: How People Make Decisions*. Copyright © 1999.

Quote courtesy Field Marshall Carver, cited in ADP AC 71940, *Land Operations*. Copyright © 2017 by British Ministry of Defence Crown.

Excerpts from William Joseph Slim, *Unofficial History*. Copyright © 1959 by Field-Marshal Sir William Slim. Reprinted 1962 by Orion Publishing Group.

Excerpts from William M. Connor, "Establishing Command Intent, A Case Study: The Encirclement of the Ruhr, March 1945" in *The Human in Command: Exploring The Modern Military Experience*. Edited by Carol McCann and Ross Pigeau. Copyright © 2000 by Kluwer Academic/Plenum Press.

Quote courtesy Ulysses S. Grant, *Memoirs and Selected Letters: Personal Memoirs of U.S. Grant, Selected Letters, 1839-1865*, vol. 2. Edited by William S. McFeely and Mary Drake McFeely. Copyright © 1990 by Literary Classics of the United States.

Quote courtesy *Genghis Khan: The Emperor of All Men*. Edited by Harold Lamb. Copyright © 1927 by Harold Lamb. Reprinted, New York: Doubleday, 1956. All rights reserved.

Quote courtesy Erwin Rommel, *The Rommel Papers*. Edited by B. H. Liddell Hart. Copyright © 1953 by B. H. Liddell Hart.

Excerpts from Robert A. Doughty, *The Breaking Point: Sedan and the Fall of France*. Copyright © 1990 by Robert A. Doughty.

Quote courtesy Richard E Simpkin and John Erickson, *Deep Battle: The Brainchild of Marshal Tukhachevskii*. Copyright © 1987 Brassey's Defence.

Quote courtesy George S. Patton, *Military Essays and Articles by George S. Patton, Jr. General, U.S. Army 02605 1885 – 1945*. Edited by Charles M. Province. Copyright © 2002 by the George S. Patton, Jr. Historical Society. All rights reserved.

Quote reprinted courtesy Erin Johnson, "Schwarzkopf Speaks of Leadership at Symposium," *The Daily Universe*, 21 October 2001. Online https://universe.byu.edu/2001/10/11/schwarzkopf-speaks-of-leadership-at-symposium/.

Quote courtesy Lt.-Col. Simonds, Commandant, "Address to Canadian Junior War Staff Course, 24 April 1941." Online https://www.canada.ca/en/department-national-defence/services/military-history/history-heritage/official-military-history-lineages/reports.html.

Introduction

This revision to ADP 6-0 represents an evolution of mission command doctrine based upon lessons learned since 2012. The use of the term *mission command* to describe multiple things—the warfighting function, the system, and a philosophy—created unforeseen ambiguity. Mission command replaced *command and control*, but in practical application it often meant the same thing. This led to differing expectations among leadership cohorts regarding the appropriate application of mission command during operations and garrison activities. Labeling multiple things mission command unintentionally eroded the importance of mission command, which is critical to the command and control of Army forces across the range of military operations. Differentiating mission command from command and control provides clarity, allows leaders to focus on mission command in the context of the missions they execute, and aligns the Army with joint and multinational partners, all of whom use the term command and control.

Command and control—the exercise of authority and direction by a properly designated commander over assigned and attached forces—is fundamental to the art and science of warfare. No single specialized military function, either by itself or combined with others, has a purpose without it. Commanders are responsible for command and control. Through command and control, commanders provide purpose and direction to integrate all military activities towards a common goal—mission accomplishment. Military operations are inherently human endeavors, characterized by violence and continuous adaptation by all participants. Successful execution requires Army forces to make and implement effective decisions faster than enemy forces. Therefore, the Army has adopted mission command as its approach to command and control that empowers subordinate decision making and decentralized execution appropriate to the situation.

The nature of operations and the patterns of military history point to the advantages of mission command. Mission command traces its roots back to the German concept of Auftragstaktik (literally, mission-type tactics). Auftragstaktik was a result of Prussian military reforms following the defeat of the Prussian army by Napoleon at the Battle of Jena in 1809. Reformers such as Gerhard von Scharnhorst, August von Gneisenau, and Helmuth von Moltke sought to develop an approach for planning campaigns and commanding large armies over extended battlefields. At the heart of the debate was a realization that subordinate commanders in the field often had a better understanding of what was happening during a battle than the general staff, and they were more likely to respond effectively to threats and fleeting opportunities if they were allowed to make decisions based on this knowledge. Subordinate commanders needed the authority to make decisions and act based on changing situations and unforeseen events not addressed in the plan. After decades of debate, professionalization of the army, practical application during the Danish-Prussian War of 1864, the Austro-Prussian War of 1866, and the Franco-Russian War of 1870, Auftragstaktik was codified in the 1888 German Drill Regulation.

In Auftragstaktik, commanders issue subordinate commanders a clearly defined goal, the resources to accomplish the goal, and a time frame to accomplish the goal. Subordinate commanders are then given the freedom to plan and execute their mission within the higher commander's intent. During execution, Auftragstaktik demanded a bias for action within the commander's intent, and it required leaders to adapt to the situation as they personally saw it, even if their decisions violated previous guidance or directives. To operate effectively under this style of command requires a common approach to operations and subordinates who are competent in their profession and trained in independent decision making.

Aspects of mission command, including commander's intent, disciplined initiative, mission orders, and mutual trust, have long been part of U.S. Army culture. The most successful U.S. Army commanders have employed elements of mission command since the 18th century. Grant's orders to Sherman for the campaign of 1864 and Sherman's supporting plan are models of clear commander's intent, mission orders, and understanding based on mutual trust. (See the vignette on page 1-9.) When addressing operations orders, the Army's 1905 *Field Service Regulation* contained the following passage that served as an early discussion of mission orders:

An order should not trespass on the province of the subordinate. It should contain everything which is beyond the independent authority of the subordinate, but nothing more. When the transmission of orders involves a considerable period of time, during which the situation may change, detailed instructions are to be avoided. The same rule holds when orders may have to be carried out under circumstances which the originator of the order cannot completely forecast; in such cases letters of guidance is more appropriate. It should lay stress upon the object to be attained, and leave open the means to be employed.

Eisenhower's general plan and intent for the 1944 invasion of Europe and defeat of Nazi Germany is an example of mission command that guided Allied forces as they fought their way from Normandy to the Rhine and beyond. A more recent example is the 3rd Infantry Division's march to Baghdad in 2003 and the subsequent "thunder runs" that showed the world that the Iraqi regime was defeated. Retired General David Perkins (a brigade commander during this operation) writes, "These thunder runs were successful because the corps and division-level commanders established clear intent in their orders and trusted their subordinates' judgment and abilities to exercise disciplined initiative in response to a fluid, complex problem, underwriting the risks that they took."

Mission command requires tactically and technically competent commanders, staffs, and subordinates operating in an environment of mutual trust and shared understanding. It requires building effective teams and a command climate in which commanders encourage subordinates to take risks and exercise disciplined initiative to seize opportunities and counter threats within the commander's intent. Through mission orders, commanders focus their subordinates on the purpose of an operation rather than on the details of how to perform assigned tasks. This allows subordinates the greatest possible freedom of action in the context of a particular situation. Finally, when delegating authority to subordinates, commanders set the necessary conditions for success by allocating resources to subordinates based on assigned tasks.

Commanders need support to exercise command and control effectively. At every echelon of command, commanders are supported by the command and control warfighting function—the related tasks and a system that enables commanders to synchronize and converge all elements of combat power. Commanders execute command and control through their staffs and subordinate leaders.

This publication provides fundamental principles on mission command, command and control, and the command and control warfighting function. Key updates and changes to this version of ADP 6-0 include—

- Combined information from ADP 6-0 and ADRP 6-0 into a single document.
- Command and control reintroduced into Army doctrine.
- An expanded discussion of command and control and its relationship to mission command.
- Revised mission command principles.
- Command and control system reintroduced, along with new tasks, and an updated system description.
- Expanded discussion of the command and control system.

ADP 6-0 contains 4 chapters:

Chapter 1 provides an overview of mission command, command, and control. It describes the nature of operations and the Army's operational concept, and how it is enabled by the mission command. It then discusses the function of command and control, and how commanders create conditions for mission command to flourish. The chapter concludes with a discussion of the command and control warfighting function.

Chapter 2 defines and describes command. It describes the nature of command, provides the elements of command, describes the role of the commander in operations, and offers guides to effective command.

Chapter 3 defines and describes control and its relationship to command. It discusses the elements of control and guides to effective control. Finally, this chapter discusses the importance of knowledge management and information management as they relate to control.

Chapter 4 discusses the command and control system that performs the functions necessary to exercise command and control. This includes a discussion of the people, processes, networks, and command posts that are components of the command and control system. It also discusses command post design and organization considerations.

Introductory table-1 lists modified terms and acronyms. The introductory figure-1 on page x illustrates the ADP 6-0 logic map.

Introductory table-1. New, modified, and removed Army terms

Term or Acronym	*Remarks*
art of command	No longer defines term.
authority	No longer defines term.
civil considerations	ADP 6-0 is now the proponent for the term and modifies the definition.
command and control	Adopts the joint definition.
command and control system	New Army definition.
commander's visualization	ADP 6-0 is now the proponent for the term.
common operational picture	Modifies the definition.
data	New definition.
essential element of friendly information	ADP 6-0 is now the proponent for the term and modifies the definition.
information	New definition.
information protection	No longer defines term.
information system	No longer defines term.
knowledge	New term and definition.
key tasks	ADP 6-0 is now the proponent for the term.
mission command	New Army definition.
mission command system	Rescinds term.
mission command warfighting function	Rescinds term.
prudent risk	Rescinds term.
relevant information	New term.
science of control	No longer defines term.
situational understanding	ADP 6-0 is now the proponent for the term.
understanding	New term and definition.

Nature of War

Military operations are inherently human endeavors representing a contest of wills, characterized by violence and continuous adaption by all participants, conducted in dynamic and uncertain operational environments to achieve a political purpose.

Operations must account for the nature of war. As such the Army's operational concept is...

Unified Land Operations

The simultaneous execution of offense, defense, stability, and defense support of civil authorities across multiple domains to shape operational environments, prevent conflict, prevail in large-scale ground combat, and consolidate gains as part of unified action.

The Army's operational concept is enabled by....

Mission Command

The Army's approach to command and control that empowers subordinate decision making and decentralized execution appropriate to the situation.

Enabled by the principles of...

Competence | Mutual trust | Shared understanding | Commander's intent
Mission orders | Disciplined initiative | Risk acceptance

Command and control is fundamental to all operations...

Command and Control

Command and control is the exercise of authority and direction by a properly designated commander over assigned and attached forces in the accomplishment of a mission.

Elements of Command
- Authority
- Responsibility
- Decision making
- Leadership

Elements of Control
- Direction
- Feedback
- Information
- Communication

Executed through...

Command and Control Warfighting Function

The related tasks and a system that enables commanders to synchronize and converge all elements of combat power.

Tasks
- Command forces
- Control operations
- Drive the operations process
- Establish the command and control system

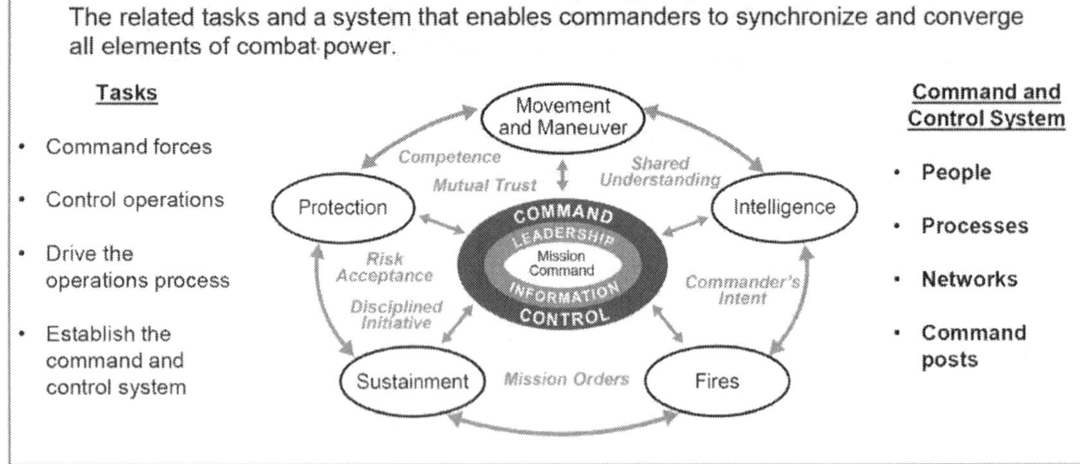

Command and Control System
- People
- Processes
- Networks
- Command posts

Introductory figure-1. Logic map

Chapter 1
Introduction to Mission Command

The situations that confront a commander in war are of infinite variety. In spite of the most careful planning and anticipation, unexpected obstacles, frictions, and mistakes are common occurrences in battle. A commander must school himself to regard these events as commonplace and not permit them to frustrate him in the accomplishment of his mission.

FM 100-5, *Operations* (1941)

This chapter sets the context for understanding mission command and command and control by describing the nature of operations and summarizing the Army's operational concept. It defines and describes mission command as the Army's approach to command and control that enables unified land operations. Then it defines and describes command and control, their relationship to each other, and their elements. The chapter concludes with a discussion of the command and control warfighting function.

THE NATURE OF OPERATIONS

1-1. Military operations fall along a competition continuum that spans cooperation to war. Between these extremes, societies maintain relationships. These relationships include economic competition, political or ideological tension, and at times armed conflict. Violent power struggles in failed states, along with the emergence of major regional powers like Russia, China, Iran, and North Korea seeking to gain strategic positions of advantage, present challenges to the joint force. Army forces must be prepared to meet these challenges across the range of military operations during periods of competition and war.

1-2. The range of military operations is a fundamental construct that helps relate military activities and operations in scope and purpose against the backdrop of the competition continuum. The potential range of military activities and operations extends from military engagement, security cooperation, and deterrence, up through large-scale combat operations in war. Whether countering terrorism as part of a limited contingency operation, or defeating a peer threat in large-scale ground combat, the nature of conflict is constant. Military operations are—

- Human endeavors.
- Conducted in dynamic and uncertain environments.
- Designed to achieve a political purpose.

HUMAN ENDEAVOR

1-3. War is a human endeavor—a clash of wills characterized by the threat or application of force and violence, often fought among populations. It is not a mechanical process that can be precisely controlled by machines, calculations, or processes. Nor is it conducted in carefully controlled and predictable environments. Fundamentally, all war is about changing human behavior. It is both a contest of wills and a contest of intellect between two or more sides in conflict, with each trying to alter the behavior of the other side. During operations, Army forces face thinking and adaptive enemies, differing agendas among the actors involved, and the variable perceptions of public opinion both inside and outside of an area of operations. As friendly forces seek to impose their will on enemy forces, the enemy forces resist and seek to impose their will on friendly forces. A similar dynamic occurs among civilian groups whose own desires influence and are influenced by military operations. All sides act, react, learn, and adapt. Appreciating these relationships is essential to understanding the fundamental nature of operations.

Dynamic and Uncertain

> *War is the realm of uncertainty; three quarters of the factors on which action in war is based are wrapped in a fog of greater or lesser uncertainty....The commander must work in a medium which his eyes cannot see; which his best deductive power cannot always fathom; and with which, because of constant changes, he can rarely become familiar.*
>
> Carl von Clausewitz

1-4. War, especially land combat, is inherently dynamic and uncertain. The complexity of friendly and enemy organizations, unique combinations of terrain and weather, and the dynamic interaction among all participants create uncertainty. Chance and friction further increase the potential for chaos and uncertainty during operations. Chance pertains to unexpected events or changes beyond the control of friendly forces, while friction describes the obstacles that make the execution of even simple tasks difficult. Both are always present for all sides during combat.

1-5. The scale, scope, tempo, and lethality of large-scale ground combat exacerbates the dynamic and uncertain nature of war, delaying or making precise cause-and effect determinations difficult. The unintended effects of operations often cannot be anticipated and may not be readily apparent. War is inherently chaotic, demanding an approach to the command and control of operations that does not attempt to impose perfect order, but rather makes allowances for uncertainty created by chance and friction.

Designed to Achieve a Political Purpose

> *[T]he role of grand strategy—higher strategy—is to co-ordinate and direct all the resources of a nation, or band of nations, towards the attainment of the political object of the war— the goal defined by fundamental policy.*
>
> B.H. Liddell-Hart

1-6. All U.S. military operations share a common fundamental purpose—to achieve specific objectives that support attainment of the overall political purpose of the operation. Objective—directing every military operation towards a clearly defined, decisive, and attainable goal—is a principle of war. In large-scale ground combat, the purpose of operations may be to destroy the enemy's capabilities and will to fight. The purpose of operations short of large-scale combat may be more nuanced and difficult to define, and they may require support to achieve multiple objectives. These operations frequently involve setting conditions that improve positions of relative advantage compared to that of a specific adversary and contribute to achieving strategic aims in an operational area without large-scale ground combat. In either case, all operations are designed to achieve the political purpose set by national authorities.

UNIFIED LAND OPERATIONS

1-7. The Army operational concept for conducting operations as part of a joint team is unified land operations. *Unified land operations* is the simultaneous execution of offense, defense, stability, and defense support of civil authorities across multiple domains to shape operational environments, prevent conflict, prevail in large-scale ground combat, and consolidate gains as part of unified action (ADP 3-0). The goal of unified land operations is to achieve the joint force commander's end state by applying landpower as part of unified action. (See ADP 3-0 for a detailed discussion of unified land operations.)

1-8. The Army's primary mission is to organize, train, and equip its forces to conduct prompt and sustained land combat to defeat enemy ground forces and seize, occupy, and defend land areas. During the conduct of unified land operations, Army forces support the joint force through four strategic roles:

- Shape operational environments.
- Prevent conflict.
- Prevail during large-scale ground combat.
- Consolidate gains.

1-9. An *operational environment* is a composite of the conditions, circumstances, and influences that affect the employment of capabilities and bear on the decisions of the commander (JP 3-0). Army forces assist in shaping an operational environment by providing trained and ready forces to geographic combatant commanders in support of their combatant commander's campaign plans. Shaping activities include security cooperation, military engagement, and forward presence to promote U.S. interests and assure allies. The theater army and subordinate Army forces assist the geographic combatant commander in building partner capacity and capability and promoting stability across an area of responsibility. Army operations to shape are continuous throughout a geographic combatant commander's area of responsibility and occur before, during, and after a specific joint operation. If operations to shape are successful, they may prevent conflict and negate the requirement to conduct large-scale ground combat operations.

1-10. Army operations to prevent conflict are designed to deter undesirable actions by an adversary through the positioning of friendly capabilities and demonstrating the will to use them. Army forces may have a significant role in the execution of directed flexible deterrent options or flexible response options. Additionally, Army prevent activities may include mobilization, force tailoring, and other pre-deployment activities; initial deployment into a theater, including echeloning command posts; and development of intelligence, communications, sustainment, and protection infrastructure to support the joint force commander.

1-11. While the Army may conduct combat operations at various levels across the range of military operations, Army forces must be manned, equipped, and trained for large-scale ground combat. During large-scale ground combat operations, Army forces focus on the defeat and destruction of enemy ground forces as part of the joint team. Army forces close with and destroy enemy forces, exploit success, and break an opponent's will to resist. Army forces attack, defend, conduct stability tasks, and consolidate gains to achieve national objectives.

1-12. Operations to consolidate gains include activities to make enduring any temporary operational success and set the conditions for a stable environment allowing for a transition of control to legitimate authorities. Army forces deliberately plan to consolidate gains during all phases of an operation. In some instances, Army forces will be in charge of integrating forces and synchronizing activities to consolidate gains. In other situations, Army forces will be in support. While Army forces consolidate gains throughout an operation, consolidating gains becomes the focus of Army forces after large-scale combat operations have concluded. (See FM 3-0 for a detailed discussion of how Army forces shape operational environments, prevent conflict, conduct large-scale ground combat, and consolidate gains.)

MISSION COMMAND

Never tell people how to do things. Tell them what to do and they will surprise you with their ingenuity.

General George S. Patton, Jr.

1-13. Army operations doctrine emphasizes shattering an enemy force's ability and will to resist, and destroying the coherence of enemy operations. Army forces accomplish these things by controlling the nature, scope, and tempo of an operation and striking simultaneously throughout the area of operations to control, neutralize, and destroy enemy forces and other objectives. The Army's command and control doctrine supports its operations doctrine. It balances coordination, personal leadership, and tactical flexibility. It stresses rapid decision making and execution, including rapid response to changing situations. It emphasizes mutual trust and shared understanding among superiors and subordinates.

1-14. ***Mission command*** **is the Army's approach to command and control that empowers subordinate decision making and decentralized execution appropriate to the situation.** Mission command supports the Army's operational concept of unified land operations and its emphasis on seizing, retaining, and exploiting the initiative.

1-15. The mission command approach to command and control is based on the Army's view that war is inherently chaotic and uncertain. No plan can account for every possibility, and most plans must change rapidly during execution to account for changes in the situation. No single person is ever sufficiently informed to make every important decision, nor can a single person keep up with the number of decisions that need to be made during combat. Subordinate leaders often have a better understanding of what is happening during a battle, and are more likely to respond effectively to threats and fleeting opportunities if allowed to make decisions and act based on changing situations and unforeseen events not addressed in the initial plan in order to achieve their commander's intent. Enemy forces may behave differently than expected, a route may become impassable, or units could consume supplies at unexpected rates. Friction and unforeseeable combinations of variables impose uncertainty in all operations and require an approach to command and control that does not attempt to impose perfect order, but rather accepts uncertainty and makes allowances for unpredictability.

1-16. Mission command helps commanders capitalize on subordinate ingenuity, innovation, and decision making to achieve the commander's intent when conditions change or current orders are no longer relevant. It requires subordinates who seek opportunities and commanders who accept risk for subordinates trying to meet their intent. Subordinate decision making and decentralized execution appropriate to the situation help manage uncertainty and enable necessary tempo at each echelon during operations. Employing the mission command approach during all garrison activities and training events is essential to creating the cultural foundation for its employment in high-risk environments.

Von Moltke and Auftragstaktik

Helmuth von Moltke (1800-1891) was appointed chief of the Prussian general staff in 1857. One of the important concepts he promulgated was Auftragstaktik (literally, "mission tactics"), a command method stressing decentralized initiative within an overall strategic design. Moltke understood that, as war progressed, its uncertainties diminished the value of any detailed planning that might have been done beforehand. He believed that, beyond calculating the initial mobilization and concentration of forces, "...no plan of operations extends with any degree of certainty beyond the first encounter with the main enemy force." He believed that, throughout a campaign, commanders had to make decisions based on a fluid, constantly evolving situation. For Moltke, each major encounter had consequences that created a new situation, which became the basis for new measures. Auftragstaktik encouraged commanders to be flexible and react immediately to changes in the situation as they developed. It replaced detailed planning with delegation of decision-making authority to subordinate commanders within the context of the higher commander's intent. Moltke realized that tactical decisions had to be made on the spot; therefore, great care was taken to encourage initiative by commanders at all levels.

Moltke believed that commanders should issue only the most essential orders. These would provide only general instructions outlining the principal objective and specific missions. Tactical details were left to subordinates. For Moltke, "the advantage which a commander thinks he can attain through continued personal intervention is largely illusory. By engaging in it he assumes a task that really belongs to others, whose effectiveness he thus destroys. He also multiplies his own tasks to a point where he can no longer fulfill the whole of them."

Moltke's thoughts, codified in the 1888 German field regulation, were imbued into the culture of the Germany Army.

Subordinate Decision Making

1-17. Successful commanders anticipate future events by developing branches and sequels instead of focusing on details better handled by subordinates during current operations. The higher the echelon, the more time commanders should devote to future operations and the broader the guidance provided to subordinates. Subordinates empowered to make decisions during operations unburden higher commanders from issues that distract from necessary broader perspective and focus on critical issues. Mission command allows those commanders with the best situational understanding to make rapid decisions without waiting for higher echelon commanders to assess the situation and issue orders.

1-18. Commanders delegate appropriate authority to deputies, subordinate commanders, and staff members based upon a judgment of their capabilities and experience. Delegation allows subordinates to decide and act for their commander in specified areas. Delegating decision-making authority reduces the number of decisions made at the higher echelons and reduces response time at lower echelons. In addition to determining the amount of decision-making authority they will delegate, commanders also identify decisions that are their sole responsibility and cannot be delegated to subordinates.

1-19. When delegating authority to subordinates, commanders strive to set the necessary conditions for success. They do this by assessing and managing risk. Taking risk is inherent at all levels of command. Commanders and staffs assess hazards and recommend controls to help manage risk, rather than forcing unnecessary risk decisions on subordinates. Risk, including ethical risk, should be identified and mitigated by the higher level commander to the greatest extent possible. Two ways of managing risk are by managing the number of tasks assigned to subordinates and by providing the appropriate resources to accomplish those tasks. These resources include information, forces, materiel, and time.

1-20. While commanders can delegate authority, they cannot delegate responsibility. Subordinates are accountable to their commanders for the use of delegated authority, but commanders remain solely responsible and accountable for the actions of their subordinates.

Decentralized Execution

1-21. Decentralized execution is the delegation of decision-making authority to subordinates, so they may make and implement decisions and adjust their assigned tasks in fluid and rapidly changing situations. Subordinate decisions should be ethically based and within the framework of their higher commander's intent. Decentralized execution is essential to seizing, retaining, and exploiting the operational initiative during operations in environments where conditions rapidly change and uncertainty is the norm. Rapidly changing situations and uncertainty are inherent in operations where commanders seek to establish a tempo and intensity that enemy forces cannot match.

1-22. Decentralized execution requires disseminating information to the lowest possible level so subordinates can make informed decisions based on a shared understanding of both the situation and their commander's intent. This empowers subordinates operating in rapidly changing conditions to exercise disciplined initiative within their commander's intent. Generally, the more dynamic the circumstances, the greater the need for initiative to make decisions at lower levels. It is the duty of subordinates to exercise initiative to achieve their commander's intent. It is the commander's responsibility to issue appropriate intent and ensure subordinates are prepared in terms of education, training, and experience to exercise initiative.

1-23. The commander's intent provides a unifying idea that allows decentralized execution within an overarching framework. It provides guidance within which individuals may exercise initiative to accomplish the desired end state. Understanding the commander's intent two echelons up further enhances unity of effort while providing the basis for decentralized decision making and execution throughout the depth of a formation. Subordinates who understand the commander's intent are far more likely to exercise initiative effectively in unexpected situations. Under the mission command approach to command and control, subordinates have both **responsibility and authority** to fulfill the commander's intent.

Levels of Control

1-24. Determining the appropriate level of control, including delegating decisions and determining how much decentralized execution to employ, is part of the art of command. The level and application of control is constantly evolving and must be continuously assessed and adjusted to ensure the level of control is appropriate to the situation. Commanders should allow subordinates the greatest freedom of action commensurate with the level of acceptable risk in a particular situation. The mission variables (mission, enemy, terrain and weather, troops and support available, time available, and civil considerations) influence how much control to impose on subordinates. Other considerations include—

- Enemy disposition and capabilities.
- Level of synchronization and integration required.
- Higher echelon headquarters constraints.
- Level of risk.

- Level of legal and ethical ambiguity.
- Rules of engagement.
- Level of unit cohesion.
- Level of training.
- Level of trust.
- Level of shared understanding.

(See figure 1-1 for a sample of the considerations for determining the appropriate level of control.)

1-25. Different operations and phases of operations may require tighter or more relaxed control over subordinate elements than other phases. Operations that require the close synchronization of multiple units, or the integration of effects in a limited amount of time, may require more detailed coordination, and be controlled in a more centralized manner. Examples of this include combined arms breaches, air assaults, and wet gap crossings. Conversely, operations that do not require the close coordination of multiple units, such as a movement to contact or a pursuit, offer many opportunities to exercise initiative. These opportunities may be lost if too much emphasis is placed on detailed synchronization. Even in a highly controlled operation, subordinates must still exercise initiative to address unexpected problems and achieve their commander's intent when existing orders no longer make sense in the context of execution.

← More control — Less control →

	Considerations	
• Predictable • Known	Situation	• Unpredictable • Unknown
• Inexperienced • New team	Unit Cohesion	• Experienced • Mature team
• Untrained or needs practice	Level of Training	• Trained in tasks to be performed
• Being developed	Level of Trust	• Established
• Top down • Explicit communications • Vertical communications	Shared Understanding	• Reciprocal information • Implicit communications • Vertical and horizontal communications
• Restrictive	Rules of Engagement	• Permissive
• Optimal decisions later	Required Decision	• Acceptable decisions sooner
• Science of war • Synchronization	Appropriate To	• Art of war • Orchestration

Figure 1-1. Levels of control

PRINCIPLES OF MISSION COMMAND

1-26. Mission command requires competent forces and an environment of mutual trust and shared understanding among commanders, staffs, and subordinates. It requires effective teams and a command climate in which subordinates are required to seize opportunities and counter threats within the commander's intent. Commanders issue mission orders that focus on the purpose of an operation and essential coordination measures rather than on the details of how to perform assigned tasks, giving subordinates the latitude to accomplish those tasks in a manner that best fits the situation. This minimizes the number of decisions a single commander makes and allows subordinates the greatest possible freedom of action to accomplish tasks. Finally, when delegating authority to subordinates, commanders set the necessary conditions for success by allocating appropriate resources to subordinates based on assigned tasks. Successful mission command is enabled by the principles of—

- Competence.
- Mutual trust.
- Shared understanding.
- Commander's intent.
- Mission orders.
- Disciplined initiative.
- Risk acceptance.

COMPETENCE

1-27. Tactically and technically competent commanders, subordinates, and teams are the basis of effective mission command. An organization's ability to operate using mission command relates directly to the competence of its Soldiers. Commanders and subordinates achieve the level of competence to perform assigned tasks to standard through training, education, assignment experience, and professional development. Commanders continually assess the competence of their subordinates and their organizations. This assessment informs the degree of trust commanders have in their subordinates' ability to execute mission orders in a decentralized fashion at acceptable levels of risk.

1-28. Training and education that occurs in both schools and units provides commanders and subordinates the experiences that allow them to achieve professional competence. Repetitive, realistic, and challenging training creates common experiences that develop the teamwork, trust, and shared understanding that commanders need to exercise mission command and forces need to achieve unity of effort. (See ADP 7-0 for doctrine on individual and collective training.)

1-29. Leaders supplement institutional and organizational training and education with continuous self-development. Self-development is particularly important for the skills that rely on the art of command, which is further developed by reading and studying the art of war. These skills can also be developed through coursework, simulations and experience. (See chapter 2 for discussion on the art of command.)

MUTUAL TRUST

1-30. Mutual trust is shared confidence between commanders, subordinates, and partners that they can be relied on and are competent in performing their assigned tasks. There are few shortcuts to gaining the trust of others. Trust is given by leaders and subordinates, and built over time based on common shared experiences. It is the result of upholding the Army values, exercising leadership consistent with Army leadership principles, and most effectively instilled by the leader's personal example.

1-31. Mutual trust is essential to successful mission command, and it must flow throughout the chain of command. Subordinates are more willing to exercise initiative when they believe their commander trusts them. They will also be more willing to exercise initiative if they believe their commander will accept and support the outcome of their decisions. Likewise, commanders delegate greater authority to subordinates who have demonstrated tactical and technical competency and whose judgment they trust.

1-32. At the lowest tactical levels the ability to trust subordinate formations to execute their collective tasks and battle drills is essential. Building that trust is critical to rapid decision making in high pressure situations; commanders should be focused more on the problem to be solved when giving guidance than the methods that their subordinates might use. Subordinates must trust that commanders will employ mission orders to the maximum extent possible once they have demonstrated the attributes and competencies expected.

1-33. Commanders must also trust their colleagues who are commanding adjacent and supporting forces, and they must earn their trust as well. When a commander exercises initiative, trust gives other commanders the same level of confidence to synchronize their actions with those of that commander. Such actions synchronize operations without requiring detailed instructions from higher echelons. Once established and sustained, trust allows each echelon to focus on operations as a whole instead of on the actions of individual subordinates.

1-34. Trust is based on personal qualities, such as professional competence, character, and commitment. Soldiers must see values in action before such actions become a basis for trust. Trust is built through shared experiences and training deliberately developed by commanders or through the conduct of operations. During shared experiences, two-way communication and interaction among the commander, subordinates, and Soldiers reinforces trust. Soldiers expect to see members of the chain of command accomplishing the mission while taking care of their welfare and leading by example through shared hardships and danger.

1-35. Trust is also a product of a common background, education, understanding of doctrine, and a common language for operations. In some situations, trust may be based solely on a common understanding of an approach to operations. This understanding creates a basic level of trust that, until proven otherwise, new team members or adjacent units will conduct operations to a common standard. During large-scale ground combat operations where task organizations are likely to change rapidly and often, commanders and staffs must assume a basic level of trust regarding the level of competence among new teams.

Shared Understanding

1-36. A critical challenge for commanders, staffs, and unified action partners is creating shared understanding of an operational environment, an operation's purpose, problems, and approaches to solving problems. *Unified action partners* are those military forces, governmental and nongovernmental organizations, and elements of the private sector with whom Army forces plan, coordinate, synchronize, and integrate during the conduct of operations (ADP 3-0). Shared understanding of the situation, along with the flow of information to the lowest possible level, forms the basis for unity of effort and subordinates' initiative. Effective decentralized execution is not possible without shared understanding.

1-37. Shared understanding starts with the Army's doctrine and professional military education that instills a common approach to the conduct of operations, a common professional language, and a common understanding of the principles of mission command. Army professionals understand the most current Army doctrine to ensure a minimum level of shared understanding for the conduct of operations. It is this shared understanding that allows even hastily task-organized units to operate effectively.

1-38. Commanders and staffs actively create shared understanding throughout the operations process (planning, preparation, execution, and assessment). They collaboratively frame an operational environment and its problems, and then they visualize approaches to solving those problems.

1-39. Collaboration is more than coordination. It is multiple people and organizations working together towards a common goal by sharing knowledge and building consensus. It requires dialogue that involves the candid exchange of ideas or opinions among participants and encourages frank discussions in areas of disagreement. Throughout the operations process, commanders, subordinate commanders, staffs, and unified action partners collaborate by sharing and questioning information, perceptions, and ideas to understand situations and make decisions.

1-40. Through collaboration, commanders create a learning environment that allows participants to think critically and creatively and share their ideas, opinions, and recommendations without fear of reproach. Effective collaboration requires candor and a free, yet mutually respectful, exchange of ideas. Participants must feel free to make viewpoints based on their expertise, experience, and insight. This includes sharing ideas that contradict the opinions held by those of higher rank. Successful commanders listen to novel ideas and counterarguments. Effective collaboration is not possible unless commanders enable it.

1-41. Commanders establish a culture of collaboration in their organization. They recognize that they do not know everything, and they recognize that they may have something to learn from even the most junior subordinate. Commanders establish a command climate by their personal example, coaching, counseling, and mentoring where collaboration routinely occurs throughout their organization. Successful commanders invest the time and effort to visit with Soldiers, subordinate leaders, and unified action partners to understand their issues and concerns. Through such interactions, subordinates and partners gain insight into their commander's leadership style and expectations.

1-42. Shared understanding both supports and derives from trust. However, like trust, it requires time to establish, and commanders cannot assume shared understanding. Shared understanding is perishable, and commanders and their subordinates must adequately communicate to maintain shared understanding of the situation, problems, and potential solutions. Commanders can develop shared understanding in their organizations through training and by creating an environment of collaboration and dialogue.

1-43. An important source of shared understanding is open and clear communications between leaders and Soldiers. Commanders can also aid shared understanding by exhibiting a demeanor and personal mannerisms that reinforce, or at least do not contradict, their spoken message. Units develop the ability to communicate through familiarity, trust, a shared philosophy, and experiences. Sharing a common perception of military problems also leads to shared understanding.

1-44. There is a hierarchical component of shared understanding. At each echelon of command, commanders will have a slightly different understanding of the situation. Having a common perception of military problems does not imply any requirement to come to identical solutions; under mission command, understanding what outcome to achieve is more important than agreement on how to achieve it. Activities that can lead to shared understanding include collaboration among commanders and staffs, professional development meetings, terrain walks, and professional discussions.

Command Based on Shared Understanding and Trust: Grant's Orders to Sherman, 1864

In a letter to MG William T. Sherman, dated 4 April 1864, LTG Ulysses S. Grant outlined his 1864 campaign plan. LTG Grant described MG Sherman's role by writing, "It is my design, if the enemy keep quiet and allow me to take the initiative in the spring campaign, to work all parts of the army together, and somewhat towards a common centre.... You I propose to move against Johnston's army, to break it up and to get into the interior of the enemy's country as far as you can, inflicting all the damage you can against their War resources. I do not propose to lay down for you a plan of campaign, but simply to lay down the work it is desirable to have done and leave you free to execute it in your own way. Submit to me, however, as early as you can, your plan of operation."

MG Sherman responded to LTG Grant immediately in a letter dated 10 April 1864. He sent LTG Grant, as requested, his specific plan of operations, demonstrating that he understood LTG Grant's intent:

"...That we are now all to act in a common plan, converging on a common center, looks like enlightened war.... I will not let side issues draw me off from your main plan, in which I am to Knock Joe [Confederate GEN Joseph E.] Johnston, and do as much damage to the resources of the enemy as possible....I would ever bear in mind that Johnston is at all times to be kept so busy that he cannot, in any event, send any part of his command against you or [Union MG Nathaniel P.] Banks."

COMMANDER'S INTENT

I suppose dozens of operation orders have gone out in my name, but I never, throughout the war, actually wrote one myself. I always had someone who could do that better than I could. One part of the order I did, however, draft myself—the intention. It is usually the shortest of all paragraphs, but it is always the most important, because it states—or it should—just what the commander intends to achieve. It is the one overriding expression of will by which everything in the order and every action by every commander and soldier in the army must be dominated. It should, therefore, be worded by the commander, himself.

Field Marshall William Joseph Slim

1-45. The *commander's intent* is a clear and concise expression of the purpose of the operation and the desired military end state that supports mission command, provides focus to the staff, and helps subordinate and supporting commanders act to achieve the commander's desired results without further orders, even when the operation does not unfold as planned (JP 3-0). The higher echelon commander's intent provides the basis for unity of effort throughout the force. Each commander's intent nests within the commander's intent two levels up. During planning, the initial commander's intent drives course of action development. During execution, the commander's intent establishes the limits within which a subordinate may exercise initiative.

1-46. The commander's intent succinctly describes what constitutes success for the operation. Commanders convey their intent in a format they determine most suitable to the situation. It may include the operation's purpose, key tasks, and conditions that define the end state. When describing the purpose of the operation, the commander's intent does not restate the "why" of the mission statement. Rather, it describes the broader purpose of the unit's operation in relationship to the higher commander's intent and concept of operations. Doing this allows subordinates to gain insight into what is expected of them, what constraints apply, and, most importantly, why the mission is being conducted. If it is longer than a brief paragraph it is probably too long.

1-47. ***Key tasks* are those significant activities the force must perform as a whole to achieve the desired end state**. Key tasks are not specified tasks for any subordinate unit; however, they may be sources of implied tasks. During execution—when significant opportunities present themselves or the concept of operations no longer fits the situation—subordinates use key tasks to keep their efforts focused on achieving the desired end state. Examples of key tasks include terrain the force must control or an effect the force must have on the enemy. Commanders include the purpose of each associated key task to facilitate subordinate decision making and disciplined initiative.

1-48. The end state is a set of desired future conditions the commander wants to exist when an operation ends. Commanders may describe the operation's end state by stating the desired conditions of the friendly force in relationship to desired conditions of the enemy, terrain, and civil considerations. A clearly defined end state promotes unity of effort among the force and with unified action partners.

1-49. The commander's intent becomes the basis on which staffs and subordinate leaders develop plans and orders. A well-crafted commander's intent conveys a clear image of an operation's purpose and desired end state. The commander's intent provides a focus for subordinates to coordinate their separate efforts. Commanders personally prepare their commander's intent. When possible, they deliver it in person. Face-to-face delivery ensures shared understanding of what the commander wants by allowing immediate clarification of specific points. Individuals can then exercise initiative within the overarching guidance provided in the commander's intent.

1-50. Commanders write and communicate their commander's intent to describe the boundaries within which subordinates may exercise initiative while maintaining unity of effort. A clear and succinct commander's intent that lower-level leaders can remember and understand, even without an order, is key to maintaining unity of effort. Soldiers two echelons down should easily remember and clearly understand the commander's intent. Commanders collaborate with subordinates to ensure they understand the commander's intent. Subordinates who understand the commander's intent are far more likely to exercise disciplined initiative in unexpected situations.

1-51. Mission command requires that subordinates use their judgment and initiative to make decisions that further their higher commander's intent. Subordinates use the commander's intent, together with the mission statement and concept of the operation, to accomplish the mission. Empowered with trust, shared understanding, and commander's intent, they can develop the situation, adapt, and act decisively in uncertain conditions.

Mission Orders

> *An order should not trespass upon the province of a subordinate. It should contain everything that the subordinate must know to carry out his mission, but nothing more... An order must be simple and understandable, being framed to suit the intelligence and*

understanding of the recipient. Above all, it must be adapted to the circumstances under which it will be received and executed.

FM 100-5, *Tentative Field Service Regulations, Operations* (1939)

1-52. An order is a communication—verbal, written, or signaled—that conveys instructions from superiors to subordinates. The five-paragraph format (situation, mission, execution, sustainment, and command and signal) is the standard for issuing Army orders. Army commanders issue orders to give guidance, assign tasks, allocate resources, and delegate authority.

1-53. Mission command requires commanders to issue mission orders. ***Mission orders*** **are directives that emphasize to subordinates the results to be attained, not how they are to achieve them**. Mission orders enable subordinates to understand the situation, their commander's mission and intent, and their own tasks. Subordinate commanders decide how to accomplish their own mission. The commander's intent and concept of operations set guidelines that provide unity of effort while allowing subordinate commanders to exercise initiative in planning, preparing, and executing their operations.

1-54. A mission order is not a separate type of order; rather, it is a technique for writing orders that allows subordinates maximum freedom of action in accomplishing missions. Mission orders should succinctly state the mission, task organization, commander's intent and concept of operations, tasks to subordinate units, and minimum essential coordinating instructions. Tasks to subordinate units include all the standard elements (who, what, when, where, and why) with particular emphasis on the purpose (why).

1-55. Mission orders should focus on the essential tasks a subordinate is to accomplish and not an extended list of every task a unit may be required to accomplish. It should never repeat items that are part of the standard operating procedures or are commonly understood by the force. Mission orders should not contain directives to a subordinate that dictate how a task is to be accomplished. That is the province of the subordinate.

1-56. Mission orders contain the proper level of detail in the context of a particular situation; they are neither so detailed that they stifle initiative nor so general that they provide insufficient direction. The proper level of detail is situationally dependent. Some phases of operations require tighter control over subordinate elements than other phases require. An air assault's air movement and landing phases, for example, require precise synchronization. Its ground maneuver plan requires less detail. As a rule, the base plan or order contains only the specific information required to provide the guidance to synchronize combat power at the decisive time and place while allowing subordinates as much freedom of action as possible. Commanders rely on subordinates' initiative and coordination to act within the commander's intent and concept of operations.

1-57. Graphics that accompany mission orders should be drawn in a manner that allows maximum flexibility during execution. They should provide enough control for those activities requiring synchronization, but they should avoid constraining subordinates' freedom of action within their areas of operation. Ideally graphics provide enough references such as checkpoints and phase lines for subordinate leaders to use them as a basis of shared understanding when deviation from the base order becomes necessary. Inherently flexible graphics and mission orders together create conditions for initiative and rapid decision making.

1-58. Using mission orders does not mean commanders do not supervise subordinates during operations. Subordinates are accountable to their commanders for the use of delegated authority, but commanders remain solely responsible and accountable for the actions over which subordinates exercise delegated authority. Thus, commanders have the responsibility to check on their subordinates and provide directions and guidance as required to focus their activities. Commanders should emphasize mission orders during training when actual consequences are low, allowing subordinates to develop their own solutions to problems, and intervening only when necessary to avoid a serious problem. This is valuable both for subordinates to gain experience in problem solving and confidence in exercising initiative and for commanders to develop an understanding of the capabilities of subordinates.

Disciplined Initiative

Every individual from the highest commander to the lowest private must always remember that inaction and neglect of opportunities will warrant more severe censure than an error of judgment in the action taken. The criterion by which a commander judges the soundness of his own decision is whether it will further the intentions of the higher commander.

FM 100-5, *Tentative Field Service Regulations, Operations* (1941)

1-59. Disciplined initiative refers to the duty individual subordinates have to exercise initiative within the constraints of the commander's intent to achieve the desired end state. Simply put, disciplined initiative is when subordinates have the discipline to follow their orders and adhere to the plan until they realize their orders and the plan are no longer suitable for the situation in which they find themselves. This may occur because the enemy does something unforeseen, there is a new or more serious threat, or a golden opportunity emerges that offers a greater chance of success than the original course of action. The subordinate leader then takes action on their own initiative to adjust to the new situation and achieve their commander's intent, reporting to the commander about the new situation when able to do so.

1-60. Leaders and subordinates who exercise disciplined initiative create opportunity by taking action to develop a situation without asking for further guidance. Commanders rely on subordinates to act to meet their intent, not simply adhere to a plan that is no longer working. A subordinate's initiative may be the starting point for seizing, retaining, and exploiting the operational initiative by forcing an enemy to respond to friendly action.

1-61. *Operational initiative* is the setting of tempo and terms of action throughout an operation (ADP 3-0). Under mission command, subordinates are required, not just permitted, to exercise disciplined initiative in the absence of orders, when current orders no longer apply, or when an opportunity or threat presents itself. The collective effect of multiple subordinates exercising disciplined initiative over time sets the conditions for friendly forces to seize the operational initiative in chaotic and ambiguous situations.

1-62. The commander's intent defines the limits within which subordinates may exercise initiative. It gives subordinates the confidence to apply their judgment in ambiguous situations because they know the mission's purpose and desired end state. They can, on their own responsibility, take actions that they think will best accomplish the mission when communication with higher echelons is intermittent or decisions must be made immediately.

1-63. When exercising initiative, neither commanders nor subordinates are independent actors. Subordinates consider at least two factors when deciding when to exercise initiative:

- Whether the benefits of the action outweigh the risk of desynchronizing the overall operation.
- Whether the action will further the higher commander's intent.

1-64. The main consideration in exercising initiative is the urgency of the situation. If time permits, subordinates attempt to communicate their new situation and recommended course of action to their commander. When subordinates communicate their intentions to their commander, their commander can assess the implications for the overall force, and for other operations, and set in motion supporting actions. However, subordinates must depart from their orders when they are unable to contact their commander or when there is a limited amount of time to seize a fleeting opportunity. If doubt exists about whether to contact their commander or depart from orders and act to seize a fleeting opportunity, subordinates should act, if they can do so within their commander's intent.

1-65. Fostering a command climate that encourages initiative requires commanders to accept risk and underwrite the good faith mistakes of subordinates in training, before the unit is committed to combat. Commanders set conditions for subordinates to learn and gain the experience they need to operate on their own. Subordinates learn to trust that they have the authority and responsibility to act, knowing their commander will back their decisions. Because mutual trust and shared understanding constitute the foundation of subordinate initiative, commanders train subordinates to act within the commander's intent in uncertain situations.

Initiative: U.S. Paratroopers in Sicily

When paratroopers assaulted Sicily during the night of July 9-10, 1943, they suffered some of the worst weather that could affect that kind of a mission. The men were supposed to conduct two airborne assaults and form a buffer zone ahead of the 7th Army's amphibious assault on the island, but winds of up to 40 knots blew them far from their intended drop zones.

The 3,400 paratroopers in the assault took heavy losses before a single pair of boots even touched the ground. But what happened next would become airborne legend. It became the story of the "The little groups of paratroopers." The little groups of paratroopers did not find cover or spend hours trying to regroup. They rucked up wherever they were and immediately began attacking everything nearby that happened to look like it belonged to the German or Italian militaries. They tore down communications lines, demolished enemy infrastructure, set up both random and planned roadblocks, ambushed Axis forces, and attacked enemy positions. A group of 16 German pillboxes that controlled key roads was neutralized despite the attacking force having only a fraction of their planned strength. This mischief had a profound effect on the defenders. The Axis forces assumed that the paratroopers were attacking in strength at each spot where a paratrooper assault was reported. So, while many little groups of paratroopers had only a few men, German estimates reported much stronger formations. The worst reports stated that there were 10 times as many attackers as were actually present. German commanders were hard-pressed to rally against what seemed to be an overwhelming attack. Some conducted limited counterattacks at what turned out to be ghosts while others remained in defensive positions or, thinking they were overrun, surrendered to American forces that were a fraction of their size.

The operation was a success, thanks in large part to the actions of little groups of paratroopers acting on their own initiative across the island until they could find a unit to form up with. Axis forces began withdrawing from the island on July 25 and Lt. Gen. George S. Patton took Messina, the last major city on Sicily, on August 17.

RISK ACCEPTANCE

Given the same amount of intelligence, timidity will do a thousand times more damage in war than audacity.

Carl von Clausewitz

1-66. In general terms, risk is the exposure of someone or something valued to danger, harm, or loss. Because risk is part of every operation, it cannot be avoided. Commanders analyze risk in collaboration with subordinates to help determine what level of risk exists and how to mitigate it. When considering how much risk to accept with a course of action, commanders consider risk to the force and risk to the mission against the perceived benefit. They apply judgment with regard to the importance of an objective, time available, and anticipated cost. Commanders need to balance the tension between protecting the force and accepting and managing risks that must be taken to accomplish their mission.

1-67. The greatest opportunity may come from the course of action with the most risk. An example of this would be committing significant forces to a potentially costly frontal attack to fix the bulk of enemy forces in place to set the conditions for their envelopment by other forces. Another would be taking a difficult but unexpected route in order to achieve surprise.

1-68. While each situation is different, commanders avoid undue caution or commitment of resources to guard against every perceived threat. An unrealistic expectation of avoiding all risk is detrimental to mission accomplishment. Waiting for perfect intelligence and synchronization may increase risk or close a window of opportunity. Mission command requires that commanders and subordinates manage accepted risk, exercise initiative, and act decisively, even when the outcome is uncertain.

1-69. Reasonably estimating and intentionally accepting risk is not gambling. Gambling is making a decision in which the commander risks the force without a reasonable level of information about the outcome. Therefore, commanders avoid gambles. Commanders carefully determine risks, analyze and minimize as many hazards as possible, and then accept risk to accomplish the mission.

THE ROLE OF SUBORDINATES IN MISSION COMMAND

1-70. The mission command approach to command and control requires active participation by personnel of all ranks and duty positions. Subordinate officers, noncommissioned officers, and Soldiers all have important roles in the exercise of mission command. During operations, subordinates are delegated authority, typically through orders and standard operating procedures, to make decisions within their commander's intent. Commanders expect subordinates to exercise this authority to further the commander's intent when changes in the situation render orders irrelevant, or when communications are lost with higher echelon headquarters.

1-71. Because mission command decentralizes decision-making authority and grants subordinates significant freedom of action, it demands more of subordinates at all levels. Commanders must train and educate subordinates so they demonstrate good judgment when exercising initiative. Subordinates must be competent in their respective fields, and they must be confident they will have the commander's support to make and implement decisions. They must embrace opportunities to assume responsibility for achieving the commander's intent.

1-72. Subordinates do not wait for a breakdown in communications or a crisis situation to learn how to act within the commander's intent. Subordinates look for every opportunity to demonstrate and exercise initiative. To the greatest extent possible, they report what they intend to do and then execute unless their commander specifically denies them permission.

1-73. As subordinates realize their commander will support sound decisions, their trust increases, and they become more willing to exercise initiative. As commanders see subordinates perform in uncertain situations, they gain trust in their subordinates' judgment and ability.

Corporal Alvin York and Mission Command

On the morning of 8 October 1918, Soldiers of the 82nd Division began an attack to sever the German supply network in the Argonne Forest. Among the men in this push was CPL Alvin York, a squad leader in the 328th Infantry Regiment.

Initially, the American attack seemed to go well as forward-deployed Germans seemingly retreated in the face of superior numbers. However, this was a ruse, as the Germans were falling back into prepared positions. Once the Americans were in the middle of the kill zone, the Germans opened fire. This was quickly followed by German artillery ripping gaping holes in the American line.

Among the first to fall was CPL York's platoon leader, LT Kirby Stewart, and as the casualties mounted, the American attack quickly foundered. With LT Stewart dead, SGT Harry Parsons assumed command of CPL York's platoon. After surveying the situation, he ordered SGT Bernard Early, CPL York, CPL Murray Savage, and CPL William Cutting to advance with their squads to a defile to the south. From here, SGT Parsons surmised that they just might be able to get behind the German lines and eliminate the machine guns that were holding up the advance.

After dodging German fire, SGT Early led his 16 men to the defilade, then up a cut in the valley that led behind the German positions. They slowly worked their way around the German infantry and subsequently surprised and captured some 70 German soldiers, which included the battalion commander. While the Americans tried to line up their prisoners, a machine-gun crew on a nearby hill yelled to the captured Germans to take cover and then opened fire. The blast of bullets killed six Americans and wounded three. CPL York was the highest ranking Soldier not hit, and he took charge of the remaining seven men.

With the surviving Americans and German prisoners clinging to the ground, CPL York seized the initiative. He charged up the hill, outflanked the German machine gun and an infantry platoon, killing 19. Seeing a large group of German reinforcements arriving from further up the hill, CPL York decided to go back to his men. As he trotted down the hill, he was spotted by a German officer who ordered a bayonet charge to kill the American. Seeing a platoon of Germans charging, CPL York slid on his side, pulled out his 1911 Colt automatic pistol, and began picking off enemy soldiers from back to front. Seeing this, the German battalion commander, who had been captured earlier, slowly got up off of the ground and approached CPL York. Standing behind CPL York, he cautiously yelled above the din, "English?" CPL York replied, "American!" In exasperation, the German commander answered, "Good Lord! If you won't shoot anymore, I will make them give up.

CPL York and his men quickly organized their prisoners, which now numbered 100, into a formation and began marching them out of the forest. During the march back to the American lines, the Americans ended up walking into another group of Germans. CPL York shrewdly secured their surrender as well, and in the end he came out with 132 prisoners. This saved his unit from destruction, thwarted a German counterattack, and allowed the 82nd Division to achieve its objective. For his heroism, CPL York was promoted to sergeant, awarded the Medal of Honor, and would go down in history as America's most celebrated hero of the First World War.

COMMAND AND CONTROL

> *If intercommunication between events in front and ideas behind are not maintained, then two battles will be fought—a mythical headquarters battle and an actual front-line one, in which case the real enemy is to be found in our own headquarters.*
>
> Major General J.F.C. Fuller

1-74. Mission command is the Army's approach to command and control. *Command and control* is the exercise of authority and direction by a properly designated commander over assigned and attached forces in the accomplishment of mission (JP 1). Command and control (also known as C2) is fundamental to the art and science of warfare. No single activity in operations is more important than command and control. Command and control by itself will not secure an objective, destroy an enemy target, or deliver supplies. Yet none of these activities could be coordinated towards a common objective, or synchronized to achieve maximum effect, without effective command and control. It is through command and control that the countless activities a military force must perform gain purpose and direction. The goal of command and control is mission accomplishment.

1-75. The focal point of command and control is the commander. Commanders assess situations, make decisions, and direct action. They provide purpose, direction, and motivation to instill the will to win. Commanders seek to understand the situation, visualize an end state and operational approach, and describe that end state and operational approach in their commander's intent and planning guidance. During execution, commanders direct the actions of subordinates and adjust operations based on changes to the situation and feedback from subordinate units, external organizations, and their staffs.

1-76. The Army's framework for organizing and putting command and control into action is the operations process. The operations process consists of the major command and control activities performed during operations (planning, preparing, executing, and continuously assessing). Commanders, supported by their staffs, employ the operations process to drive the conceptual and detailed planning necessary to understand, visualize, and describe their operational environment; make and articulate decisions; and direct, lead, and assess military operations. (See chapter 2 for more information on the role of the commander. See ADP 5-0 for details on conducting the operations process.)

RELATIONSHIP BETWEEN COMMAND AND CONTROL

1-77. Command and control are interrelated. Command resides with commanders and includes the authority and responsibility for effectively using available resources and for planning the employment of, organizing, directing, coordinating, and controlling military forces for the accomplishment of missions. It also includes responsibility for the health, welfare, morale, and discipline of assigned personnel. Command emphasizes a commander's lawful assignment of authority and the responsibility that accompanies that authority.

1-78. Effective command is impossible without control. Control is inherent in command and includes collecting, processing, displaying, storing and disseminating relevant information. Commanders, supported by their staffs, control operations by receiving and communicating information to build shared understanding and to direct, coordinate, and synchronize the actions of subordinate units. Commander's intent, orders, control measures, and standard operating procedures all assist with the control of operations. Determining the appropriate level of control in a particular situation is a critical command responsibility.

1-79. Command and control is not a one-way, top-down process. In application, command and control is multidirectional, with feedback from lower echelons, from higher echelons, laterally, and from sources outside the chain of command. It includes the reciprocal flow of information between commanders, staffs, subordinates, and other organizations in an area of operations as they work to achieve shared understanding and adjust to continuously changing circumstances in an operational environment.

Command

> *I believe firmly in a "personal" command, i.e. that a commander should never attempt to control an operation or a battle by remaining at his H.Q. or be content to keep touch with his subordinates by cable, [radio], or other means of communication. He must as far as possible see the ground for himself to confirm or correct his impressions of the map; his*

subordinate commanders to discuss their plans and ideas with them; and the troops to judge of their needs and their morale.

Field-Marshall Earl Wavell

1-80. *Command* is the authority that a commander in the armed forces lawfully exercises over subordinates by virtue of rank or assignment (JP 1). Command is personal—an individual person commands, not an organization or a headquarters. Command is considered more art than science, because it incorporates intangible elements that require judgment in application. The key elements of command are—

- Authority.
- Responsibility.
- Decision making.
- Leadership.

1-81. Inherent in command is the authority that a military commander lawfully exercises over subordinates, including the authority to assign tasks and the responsibility for their successful completion. Authority derives from two sources: official authority and personal authority. Official authority is a function of rank and position and is bestowed by law and regulation. This authority to enforce orders by law is one of the key elements of command and distinguishes military commanders from other leaders and managers. Personal authority is a function of personal influence and derives from factors such as experience, reputation, skill, character, and personal example. It is bestowed by the other members of the organization. Official authority provides the power to act, but it is rarely enough for success on its own; the most effective commanders also possess a high degree of personal authority.

1-82. With authority comes responsibility. Commanders are legally and ethically responsible for their decisions and for the actions, accomplishments, and failures of their subordinates. Commanders may delegate authority, but delegation does not absolve commanders of their responsibility to the higher echelon commander. Commanders are always accountable for what happens or fails to happen in their command.

1-83. Commanders exercise their authority through decision making and leadership. Decision making refers to selecting a course of action as the one most favorable to accomplish the mission, and includes making adjustments to plans during the execution of an operation. Decision making includes knowing whether to decide or not, then when and what to decide, and finally, understanding the consequences. Commanders use understanding, visualization, description, and direction to make and communicate their decisions. Commanders rely on their education, experience, knowledge, and judgment in applying authority as they decide (plan how to achieve the end state) and lead (direct their forces during preparation and execution), all while assessing progress.

1-84. Leadership refers to influencing people by providing purpose, direction, and motivation, while operating to accomplish the mission and improve the organization. It is the unifying and multiplying element of combat power. Commanders lead through a combination of personal example, persuasion, and compulsion. Commanders employ leadership to translate decisions into effective action by forces.

Control

1-85. ***Control* is the regulation of forces and warfighting functions to accomplish the mission in accordance with the commander's intent.** Commanders exercise control over all forces in their area of operations, aided by their command and control system. Staffs aid the commander in exercising control by supporting the commander's decision making; assisting subordinate commanders, staffs and units; and keeping units and organizations outside the headquarters informed.

1-86. Control is more science than art because it relies on objectivity, empirical methods, and analysis. Control demands commanders and staffs understand those aspects of operations that they can analyze and measure. These include the physical capabilities and limitations of friendly and enemy organizations and systems. Control also requires a realistic appreciation for time and distance factors and the time required to initiate and complete certain actions. Units are bound by factors such as movement rates, supply consumption, weapons effects, and ethical and legal considerations.

1-87. The proper application of control incorporates some level of art, since commanders must use judgment with regard to the abilities of subordinates and the likelihood that friction is part of every operation. The key elements of control are—

- Direction.
- Feedback.
- Information.
- Communication.

1-88. Commanders, assisted by their staffs, direct the actions of their subordinates within their commander's intent, the unit's mission, and the concept of operations. Commanders provide direction and communicate information, usually in plans and orders that provide subordinate forces their tasks and instruct them how to cooperate within a broader concept of operations. In the process, they receive feedback from subordinates and supporting forces that allows commanders to update their visualization and modify plans. This feedback creates a reciprocal flow of information that leads to a shared understanding among all participants.

1-89. Central to providing direction and receiving feedback is information. The amount of information that is available makes managing information and turning it into effective decisions and actions critical to success during operations. Commanders and staffs employ knowledge management techniques to add clarity to information, speed its dissemination, and support situational understanding and decision making. (See chapter 3 for more information on knowledge management and information management)

1-90. Commanders and staffs disseminate information among people, elements, and places. Communication is more than mere transmission of information. It is an activity that allows commanders, subordinates, and unified action partners to create shared understanding that supports action. It is a means to exercise control over forces. Communication links information to decisions and decisions to action. Communication among the parts of a command enables their coordinated action. Effective commanders understand the importance of using multiple means of communication to ensure shared understanding. They use multidirectional communication and suitable means of collaboration to ensure clear understanding of the commander's intent. They also anticipate those times when communication is likely to be intermittent, and they adjust their level of control accordingly.

Command and Control in Multinational Environments

1-91. *Multinational operations* is a collective term to describe military actions conducted by forces of two or more nations, usually undertaken within the structure of the coalition or alliance (JP 3-16). Multinational operations are driven by common agreement among the participating alliance or coalition partners. While each nation has its own interests and often participates within the limitations of national caveats, all nations bring value to an operation. Each nation's force has unique capabilities, and each usually contributes to the operation's legitimacy in terms of international or local acceptability. Army forces should anticipate that most operations will be multinational operations and plan accordingly.

1-92. Multinational operations present unique command and control challenges. These include cultural and language issues, interoperability challenges, national caveats on the use of respective forces, the sharing of information and intelligence, and rules of engagement. Establishing standard operating procedures and liaison with multinational partners is critical to effective command and control. When commanding and controlling forces within a multinational training or operational setting, Army commanders should be familiar with and employ multinational doctrine and standards ratified by the United States. For example, *Allied Tactical Publication* 3.2.2 applies to Army forces during the conduct of North Atlantic Treaty Organization operations. (See FM 3-16 for a detailed discussion on multinational operations.)

Domain Command and Control Considerations

1-93. While Army commanders are primarily concerned with command and control in the land domain, command and control occurs in all domains across the range of military operations. Through command and control, Army forces converge effects from all domains (land, air, maritime, space, cyberspace), as well as the information environment and the electromagnetic spectrum, to accomplish missions. Each domain has a unique set of characteristics that influences how capabilities are synchronized and converged throughout an operation. Convergence is the continuous integration of capabilities from multiple domains, the electromagnetic spectrum, and the information environment, to create multiple dilemmas for the enemy. In order to effectively converge effects from all domains, Army forces must understand the authorities, processes, procedures, and time it takes to receive and assess effects from other domains and for Army forces to create effects in those domains.

1-94. Army forces have effectively integrated capabilities and synchronized actions in the land, air, and maritime domains for decades, and the authorities, processes, and procedures are well established. However, the military application of technologies to the space and cyberspace domains, as well as the information environment and the electromagnetic spectrum, require special consideration in planning to converge effects within or across domains. Liaisons can assist commanders with integrating capabilities resident in domains other than land. (See chapter 4 for a discussion on liaisons.)

THE COMMAND AND CONTROL WARFIGHTING FUNCTION

1-95. A *warfighting function* is a group of tasks and systems united by a common purpose that commanders use to accomplish missions and training objectives (ADP 3-0). Warfighting functions are the physical means that tactical commanders use to execute operations and accomplish missions assigned by higher level commanders. The purpose of warfighting functions is to provide an intellectual organization for common critical capabilities available to commanders and staffs at all echelons.

1-96. Operations executed through simultaneous offensive, defensive, stability, or defense support of civil authorities operations require the continuous generation and application of combat power. *Combat power* is the total means of destructive, constructive, and information capabilities that a military unit or formation can apply at one time (ADP 3-0). Combat power includes all capabilities provided by unified action partners that are integrated and synchronized with the commander's objectives to achieve unity of effort in sustained operations.

1-97. Combat power has eight elements: leadership, information, command and control, movement and maneuver, intelligence, fires, sustainment, and protection. The elements facilitate Army forces accessing joint and multinational fires and assets. The Army collectively describes the last six elements as warfighting functions. Commanders apply combat power through the warfighting functions using leadership and information. Leadership is a multiplying and unifying element of combat power. Information enables commanders at all levels to make informed decisions about the application of combat power and achieve definitive results.

1-98. The *command and control warfighting function* is the related tasks and a system that enable commanders to synchronize and converge all elements of combat power (ADP 3-0). The primary purpose of the command and control warfighting function is to assist commanders in integrating the other elements of combat power (movement and maneuver, intelligence, fires, sustainment, protection, information and leadership) to achieve objectives and accomplish missions. The command and control warfighting function consists of the command and control warfighting function tasks and the command and control system, as depicted in figure 1-2 on page 1-20.

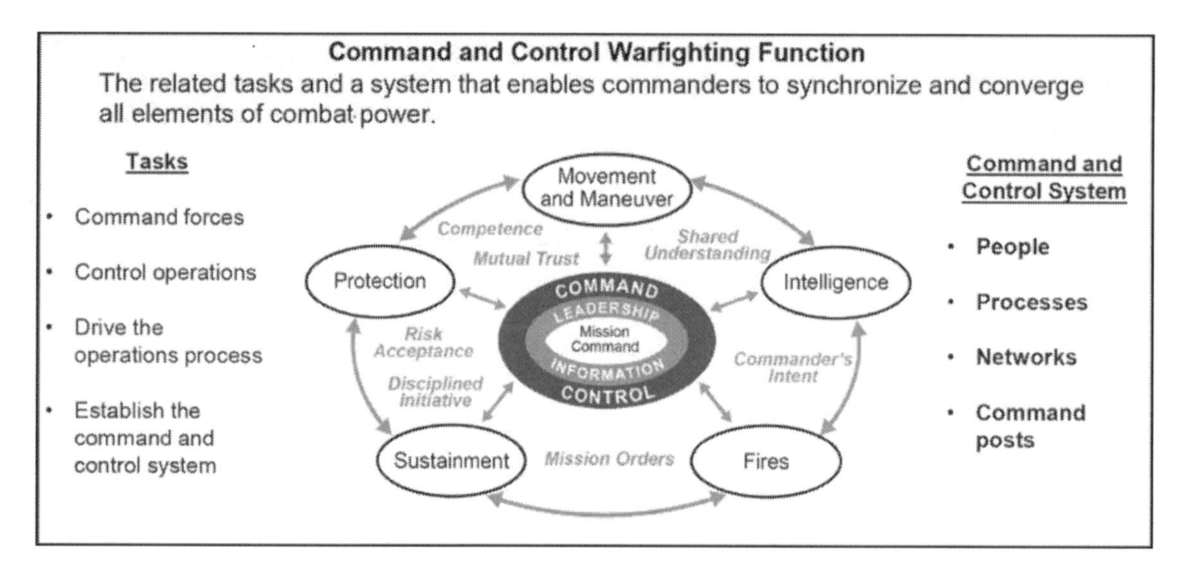

Figure 1-2. Combat power model

Command and Control Warfighting Function Tasks

1-99. The command and control warfighting function tasks focus on integrating the activities of the other elements of combat power to accomplish missions. Commanders, assisted by their staffs, integrate numerous processes and activities within their headquarters and across the force through the mission command warfighting function. These tasks are—

- Command forces (described in chapter 2).
- Control operations (described in chapter 3).
- Drive the operations process (described in chapter 2).
- Establish the command and control system (described in chapter 4).

Command and Control System

1-100. Commanders need support to effectively exercise command and control. At every echelon of command, each commander establishes a ***command and control system*****—the arrangement of people, processes, networks, and command posts that enable commanders to conduct operations.** The command and control system supports the commander's decision making, disseminates the commander's decisions to subordinates, and facilitates controlling forces. Commanders employ their command and control system to enable the people and formations conducting operations to work towards a common purpose. All the equipment and procedures exist to achieve this end. Commanders organize the four components of their command and control system to support decision making and facilitate communication. The most important of these components is people. (See chapter 4 for a detailed discussion of the command and control system.)

People

1-101. A commander's command and control system is based on people. The human aspects of operations remain paramount regardless of the technology associated with the system. Therefore, commanders base their command and control systems on human characteristics more than on equipment and processes. Trained personnel are essential to an effective command and control system. Technology cannot support command and control without them.

Processes

1-102. Commanders establish and use processes and procedures to organize activities within their headquarters and throughout the force. A process is a series of actions or steps taken to achieve a specific end, such as the military decision-making process. In addition to the major activities of the operations process, commanders and staffs use several integrating processes to synchronize specific functions throughout the operations process. The integrating processes are—

- Intelligence preparation of the battlefield (described in ATP 2-01.3).
- Information collection (described in FM 3-55).
- Targeting (described in ATP 3-60).
- Risk management (described in ATP 5-19).
- Knowledge management (See ATP 6-01.1)

1-103. *Procedures* are standard, detailed steps that prescribe how to perform specific tasks (CJCSM 5120.01). Procedures govern actions within the command and control system to make it more effective and efficient. For example, standard operating procedures often provide detailed unit instructions on how to configure common operational picture (COP) displays. Adhering to processes and procedures minimizes confusion, misunderstanding, and hesitation as commanders frequently make rapid decisions to meet operational requirements.

1-104. Processes and procedures can increase organizational competence, for example, by improving a staff's efficiency or by increasing the tempo. Processes and procedures can be especially useful in improving the coordination of Soldiers who must cooperate to accomplish repetitive tasks, such as those needed for the internal functioning of a command post. Units avoid applying procedures blindly to the wrong tasks or the wrong situations, which can lead to ineffective, even counterproductive, performance.

Networks

1-105. Generally, a network is a grouping of things that are interconnected for a purpose. Networks enable commanders to communicate information and control forces. Networks enable successful operations. Commanders determine their information requirements and focus their staffs and organizations on using networks to meet these requirements. These capabilities relieve staffs from handling routine data, and they enable extensive information sharing, collaborative planning, execution, and assessment that promote shared understanding. Each network consists of—

- End-user applications.
- Information services and data.
- Network transport and management.

Command Posts

1-106. Command posts provide a physical location for the other three components of a command and control system (people, processes, and networks). Command posts vary in size, complexity and focus. Command posts may be comprised of vehicles, containers, and tents, or located in buildings. Commanders systematically arrange platforms, operation centers, signal nodes, and support equipment in ways best suited for a particular operational environment. (See FM 6-0 and ATP 6-0.5 for more information on command posts.)

CONCLUSION

1-107. Command and control is fundamental to all operations. Mission command—the Army's approach to command and control—underpins how the U.S. Army fights. Mission command concentrates on the objective of an operation, not on every task necessary to achieve it. Mission command emphasizes timely decision making, understanding of the higher commander's intent, and the clear responsibility of subordinates to exercise initiative within that intent to achieve the desired end state. Mission command relies on decentralized execution and subordinate initiative within the commander's intent to provide unity of effort.

1-108. The fundamental basis of mission command is tactically and technically competent commanders and subordinates with a shared understanding of purpose who can be trusted to make ethical and effective decisions in the absence of further guidance. This allows commanders to focus on their intent, and it allows staffs to generate the mission orders essential to decentralized execution. Subordinates are empowered to decide when to adapt their assigned tasks to achieve the overall purpose of an operation. This requires commanders to accept risk on their subordinates' behalf and subordinates to assume responsibility for the initiative necessary for success.

1-109. In practice, mission command tends to be decentralized, informal, and flexible. Plans, orders, and graphics should be as simple and concise as possible, designed for maximum flexibility during execution. By decentralizing decision-making authority, mission command increases tempo and improves subordinates' abilities to act quickly in fluid and chaotic situations.

1-110. Commanders cannot exercise command and control alone except in the simplest and smallest of units. Even at the lowest level, commanders need support to exercise command and control effectively. At every echelon of command, each commander has a command and control system. Commanders arrange people, processes, and networks into command posts to best facilitate their exercise of authority and direction to accomplish the mission.

Chapter 2
Command

When you are commanding, leading [soldiers] under conditions where physical exhaustion and privations must be ignored, where the lives of [soldiers] may be sacrificed, then, the efficiency of your leadership will depend only to a minor degree on your tactical ability. It will primarily be determined by your character, your reputation, not much for courage—which will be accepted as a matter of course—but by the previous reputation you have established for fairness, for that high minded patriotic purpose, that quality of unswerving determination to carry through any military task assigned to you.

General of the Army George C. Marshall
Speaking to officer candidates in September 1941

This chapter begins with a discussion of the nature and elements of command. It then describes the role of the commander in operations. The chapter concludes with a discussion of guides to effective command.

NATURE OF COMMAND

2-1. Command is the authority that a commander in the armed forces lawfully exercises over subordinates by virtue of rank or assignment. Command includes the authority and responsibility for effectively using available resources and for planning the employment of, organizing, directing, coordinating, and controlling military forces for the accomplishment of missions. It also includes responsibility for health, welfare, morale, and discipline of assigned personnel.

2-2. Command is personal. In Army regulations and doctrine, an individual is given the authority to command, not an institution or group. The legal and ethical responsibilities of a commander exceed those of any other leader of similar rank who is serving in a staff position. The commander alone is responsible for what the command does or fails to do.

2-3. Command is more art than science because it requires judgment and depends on actions only human beings can perform. The art of command comprises the creative and skillful exercise of authority through timely decision making and leadership. Commanders constantly use judgment gained from experience and training to delegate authority, make decisions, determine the appropriate degree of control, and allocate resources. Proficiency in the art of command stems from years of schooling, self-development, introspection, and operational and training experiences. It also requires a deep understanding of the science of war.

2-4. The basic techniques of command do not change between echelons. However, direct leadership within a command decreases as the level of command increases, and applying organizational leadership as described in ADP 6-22 becomes more relevant. Officers prepare for higher command by developing and exercising their skills when commanding at lower levels.

ELEMENTS OF COMMAND

2-5. The elements of command are authority, responsibility, decision making, and leadership. The definition of command refers explicitly to authority. With authority comes the responsibility to carry forward an assigned task to a successful conclusion. Commanders exercise their authority by making decisions and leading their command in implementation of those decisions. Successful commanders develop skill in each element through maturity, experience, and education.

AUTHORITY

2-6. Authority is the right and power to judge, act, or command. Legal authority to enforce orders under the law is a key aspect of command and distinguishes military commanders from staff officers, and civilian leaders. Commanders understand that operations affect and are affected by human interactions. As such, they seek to establish personal authority. A commander's personal authority reinforces that commander's legal authority. Personal authority ultimately arises from the actions of commanders and the resulting trust and confidence generated by these actions. Commanders earn respect and trust by upholding laws, adhering to the Army ethic, applying Army leadership principles, and demonstrating tactical and technical expertise.

Assuming Command: General Ridgway Takes Eighth Army

After secretly crossing the Yalu River in mid-October 1950, the Chinese 13th Army Group launched its first offensive against United Nations forces on 25 October. By the end of November, the 13th Army Group had pushed most of the Eighth U.S. Army out of northwest Korea. By mid-December, the Eighth Army had retreated south of the 38th Parallel. To add to the United Nations forces' troubles, the Eighth Army commander, GEN Walton Walker, was killed in a jeep accident on 23 December.

Back in Virginia while visiting friends before Christmas, then Deputy Chief of Staff for Operations, GEN Matthew Ridgway, recalls the phone call he received from the Army Chief of Staff, GEN Joe Collins. "Matt, "he said, "I'm sorry to tell you that Johnny Walker has been killed in a jeep accident in Korea. I want you to get your things together and get out there just as soon as you can."

In his book, *Soldier: The Memoirs of Matthew B. Ridgway*, GEN Ridgway describes his initial thoughts on assuming command of the Eighth Army:

"Quickly I ran over in my mind all that I knew of the situation. As Deputy Chief of Staff for Operations, the map of Korea had become as familiar to me as the lines in my hand. I knew our strength, and our weakness. I knew personally all the top commanders in Eighth Army, except General Oliver Smith of the 1st Marine Division, and from what I knew of him, I knew I could depend on him implicitly.

Armed with this background knowledge, what should I do? Quickly, a pattern of action took place in my mind. First, of course, I would report to General MacArthur, and receive from him his estimate of the situation and broad general directives concerning operations. Next, I would assume command—and this, I knew, must be done in one simple, brief, sincere statement which would convey to Eighth Army my supreme confidence that it could turn and face and fight and defeat the Chinese horde that had sprung so suddenly from beyond the Yalu. Once this was done, I would meet with Eighth Army staff and get from them their appraisal of the situation. After that, I would call on every commander in his battle area, look into his face and the faces of his men, and form my own opinion of his firmness and resolution—or the lack thereof. Once these things were done, then I could begin to plan, could make the big decision whether to stand and hold, or to attack."

Under GEN Ridgway's leadership, the Chinese offensive was slowed and finally brought to a halt at the battles of Chipyong-ni and Wonju. He then led his troops in Operation Thunderbolt, a counter-offensive in early 1951.

2-7. Commanders may delegate authority to subordinates to accomplish a mission or assist in fulfilling their responsibilities. This includes delegating authority to members of their staffs. Delegation allows subordinates to decide and act for their commander, or in their commander's name, in specified areas. When delegating authority, commanders still remain accountable to their superiors for mission accomplishment, for the lives and care of their Soldiers, and for effectively using Army resources. Therefore, commanders use judgment in determining how much authority to delegate.

RESPONSIBILITY

2-8. Commanders are legally and ethically accountable for the decisions they make or do not make, and for the actions, accomplishments, and failures of their subordinates. Commanders may delegate authority, but they still retain overall accountability for the actions of their commands.

2-9. Command responsibilities include mission accomplishment; the health, welfare, morale, and discipline of Soldiers; and the use and maintenance of resources. In most cases, these responsibilities do not conflict; however, the responsibility for mission accomplishment sometimes conflicts with the responsibility for Soldiers or the stewardship of resources. The importance of the mission informs commanders how much risk to Soldiers and equipment to accept. When there is conflict among the three, mission accomplishment comes before Soldiers, and Soldiers come before concerns for resources. Commanders try to keep such conflicts to an absolute minimum.

DECISION MAKING

> *The commander must permit freedom of action to his subordinates insofar that this does not endanger the whole scheme. He must not surrender to them those decisions for which he alone is responsible.*
>
> German Field Service Regulation, *Truppen Fuhrung* (1935)

2-10. Decision making involves applying both the art and science of war. Many aspects of military operations—movement rates, fuels consumption, weapons effects—can be reduced to numbers, calculations, and tables. They belong to the science of war and are important to understanding what is possible with the resources available. Other aspects—the impact of leadership, complexity of operations, and uncertainty about the enemy—belong to the art of war. Successful commanders focus the most attention on those aspects belonging to the art of war.

2-11. For Army forces, decision making focuses on selecting a course of action that is most favorable to accomplishing the mission. Decision making can be deliberate, using the military decision-making process and a full staff, or decision making can be done very quickly by the commander alone. A commander's decisions ultimately guide the actions of the force.

2-12. Decision making requires knowing if, when, and what to decide as well as understanding the consequences of that decision. Critical to decision making is the ability to make decisions without perfect information, knowing when enough information allows acceptable decisions, and the willingness to act on imperfect information. Striking the balance between acting now with imperfect information and acting later with better information is essential to the art of command.

Understanding

2-13. Success in operations demands timely and effective decisions based on applying judgment to available information and knowledge. As such, commanders and staffs seek to build and maintain situational understanding throughout an operation. ***Situational understanding* is the product of applying analysis and judgment to relevant information to determine the relationships among the operational and mission variables.** Situational understanding allows commanders to make effective decisions and regulate the actions of their force with plans appropriate for the situation. It enables commanders and staffs to assess operations accurately. Commanders and staffs continually strive to maintain their situational understanding and work through periods of reduced understanding as a situation evolves. Effective commanders accept that uncertainty can never be eliminated, and they train their staffs and subordinates to function in uncertain environments.

2-14. Knowledge management and information management assist commanders with progressively adding meaning at each level of processing and analyzing to help build and maintain their situational understanding. Knowledge management and information management are interrelated activities that support commanders' decision making. There are four levels of meaning. From the lowest level to the highest level, they include data, information, knowledge, and understanding. At the lowest level, processing transforms data into information. Analysis then refines information into knowledge. Commanders and staffs then apply judgment to transform knowledge into understanding. Commanders and staffs continue a progressive development of learning, as organizations and individuals assign meaning and value at each level. (See figure 2-1.)

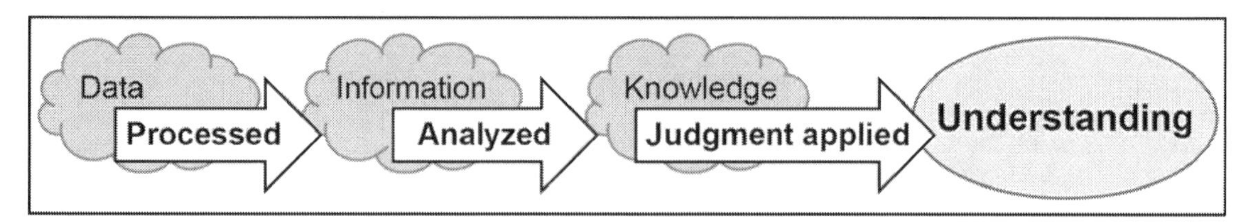

Figure 2-1. Achieving understanding

2-15. **In the context of decision making, *data* consists of unprocessed observations detected by a collector of any kind (human, mechanical, or electronic).** In typical organizations, data often flows to command posts from subordinate units. Subordinate units push data to inform higher headquarters of events that facilitate situational understanding. Data can be quantified, stored, and organized in files and databases; however, data only becomes useful when processed into information.

2-16. **In the context of decision making, *information* is data that has been organized and processed in order to provide context for further analysis**. The amount of information that is available makes managing information and using it to make effective decisions critical to success during operations. Commanders and staffs apply experience and judgment to information to gain shared understanding of events and conditions in which they make decisions during operations. Effective command and control requires further developing information into knowledge so commanders can achieve understanding

2-17. **In the context of decision making, *knowledge* is information that has been analyzed and evaluated for operational implications.** It is also comprehension gained through study, experience, practice, and human interaction that provides the basis for expertise and skilled judgment. Staffs work to improve and share tacit and explicit knowledge.

2-18. Tacit knowledge resides in an individual's mind. It is the purview of individuals, not technology. All individuals have a unique, personal store of knowledge gained from experience, training, and other people. This knowledge includes an appreciation for nuances, subtleties, and work-arounds. Intuition, mental agility, effective responses to crises, and the ability to adapt are forms of tacit knowledge. Leaders use tacit knowledge, their own and that of their subordinates, to solve complex problems and make decisions.

2-19. Explicit knowledge consists of information that can be organized, applied, and transferred using digital (such as computer files) or non-digital (such as paper) means. Explicit knowledge lends itself to rules, limits, and precise meanings. Examples of explicit knowledge include doctrinal publications, orders, and databases. Explicit knowledge is primarily used to support situational awareness and shared understanding as it applies to decision making.

2-20. **In the context of decision making, *understanding* is knowledge that has been synthesized and had judgment applied to comprehend the situation's inner relationships, enable decision making, and drive action.** Understanding is judgment applied to knowledge in the context of a particular situation. Understanding is knowing enough about the situation to change it by applying action. Judgment is based on experience, expertise, and intuition. Ideally, true understanding should be the basis for decisions. However, uncertainty and time preclude achieving perfect understanding before deciding and acting. (See chapter 3 for more information on knowledge management and information management.)

Critical and Creative Thinking

2-21. Commanders and staffs apply critical and creative thinking, including ethical reasoning, to decision making. Critical thinking examines a problem in depth from multiple points of view. It determines whether adequate justification exists to accept conclusions as true based on a given inference or argument. Critical thinkers apply judgment about what to believe or what to do in response to facts, experience, or arguments.

2-22. Creative thinking involves thinking in new, innovative ways using imagination, insight, and different ideas. Leaders often face unfamiliar problems or old problems requiring new solutions. Even situations that appear familiar require creative solutions, since an enemy force will adapt to past friendly approaches. Leaders look at different options to solve problems using lessons from similar circumstances in the past, as well as innovative approaches that come from new ideas. Creatively combining old and new ideas can create difficult dilemmas for an enemy force.

2-23. Commanders choose a decision-making approach based on the situation. There are generally two approaches to making decisions: analytic and intuitive. In certain situations, commanders may take a more deliberate approach, using systematic analysis. In other situations, commanders may rely heavily on intuition. Effective commanders consider their experience, their staff's experience, and the time and information available when considering their decision-making approach.

Analytic Decision Making

2-24. Analytic decision making generates several alternative solutions, compares those solutions to a set of criteria, and selects the best course of action. It aims to produce the optimal solution by comparing options. It emphasizes analytic reasoning guided by experience, and commanders use it when time is available. This approach offers several advantages. Analytic decision making—

- Is methodical and allows the breakdown of tasks into recognizable elements.
- Ensures commanders consider, analyze, and evaluate relevant factors and employ techniques such as war-gaming.
- Provides a systematic approach when a decision involves processing large amounts of information.
- Helps resolve conflicts among courses of action.
- Gives inexperienced personnel a logically structured approach.

2-25. Analytic decision making presents some disadvantages. It is often time-consuming, relies on large amounts of information, and requires clearly established evaluation criteria. While it is methodical, changes in conditions may require a complete reevaluation, which could delay decisions. When using this approach, commanders weigh the need for analysis against time considerations. Analytic decision making is not appropriate for all situations, especially during execution, when forces must adapt to rapidly changing situations. (An example of analytic decision making is the military decision-making process found in FM 6-0.)

Intuitive Decision Making

> *Intuition depends on the use of experience to recognize key patterns that indicate the dynamics of the situation. Because patterns can be subtle, people often cannot describe what they noticed, or how they judged a situation as typical or atypical.*
>
> Gary Klein, *Sources of Power*

2-26. Intuitive decision making is reaching a conclusion in a way that is not expressly known by the decision maker. It normally involves pattern recognition based on knowledge, judgment, experience, education, intelligence, boldness, perception, and character. Intuitive decision making—

- Focuses on assessment of the situation more than on comparing multiple options.
- Usually results in a decision for an acceptable course of action instead of the optimal course of action derived from an analytical approach.
- May be more effective when time is short.
- Relies on a commander's experience and ability to recognize the key elements and implications of a particular problem or situation.
- Tends to focus on the larger picture more than on individual components.

2-27. Intuitive decision making is faster than analytic decision making, but it requires an adequate level of experience to recognize an acceptable course of action. There is a difference between sound intuition and uninformed assumptions. When using intuitive decision making, leaders should be aware of their own biases and the differences between their current operational environment and those they experienced in the past.

2-28. By emphasizing experienced judgment over deliberate analysis, commanders increase tempo and retain flexibility to deal with uncertainty. The intuitive approach is consistent with the fact that there are no perfect solutions to battlefield problems. Even when making intuitive decisions, commanders may have time available to employ more analytic techniques. Time permitting, commanders use their staffs to validate intuitive decisions to ensure they are feasible, acceptable, and suitable.

2-29. In practice, the two approaches are rarely mutually exclusive. When time is not critical, commanders use an analytical approach or incorporate analysis into their intuitive decisions. Commanders blend intuitive and analytic decision making to help them remain objective and make timely, effective decisions. Commanders avoid making decisions purely by intuition and incorporate analysis into their intuitive decisions. Commanders seek as much analysis as possible within the time available. In situations requiring immediate decisions, this analysis may be no more than the commander's own rapid mental development, analysis, and selection of a course of action—which may appear completely intuitive. Combining both approaches best accounts for the many factors that affect decisions. An example of an approach that combines analytic and intuitive decision making is the rapid decision-making and synchronization process found in FM 6-0.

Judgment

> *Despite the years of thought and oceans of ink which have been devoted to the elucidation of war its secrets still remain shrouded in mystery....War is an art and as such is not susceptible of explanation by fixed formulae. Yet from the earliest time there has been an unending effort to subject its complex and emotional structure to dissection, to enunciate rules for its waging, to make tangible its intangibility.*
>
> General George S. Patton, Jr.

2-30. Commanders make decisions using judgment acquired from experience, training, and study. Experience contributes to judgment by providing a basis for rapidly identifying practical courses of actions and dismissing impractical ones. Commanders use judgment to assess information, situations, or circumstances shrewdly and draw feasible conclusions. Skilled judgment helps commanders form sound opinions and make sensible decisions.

2-31. Judgment is required for selecting the critical time and place to act. Commanders act by assigning missions, prioritizing, managing risk, allocating resources, and leading. Thorough knowledge of the science of war, a strong ethical sense, and an understanding of enemy and friendly capabilities form the basis of the judgment commanders require.

2-32. Judgment becomes more refined as commanders become more experienced. Increasing their knowledge, developing their intellect, and gaining experience allow commanders to develop the greater judgment required for increased responsibilities. In addition to decision-making, commanders apply their judgment to—

- Identify, accept, and mitigate risk.
- Delegate authority.
- Prioritize resources.
- Direct the staff.

Identify, Mitigate, and Accept Risk

2-33. Commanders use judgment when identifying risk by deciding how much risk to accept, mitigating risk where possible, and managing the risk they must accept. They accept risk when seizing opportunities. They reduce risk with foresight and planning, while regularly examining any assumptions associated with previous risk-related decisions.

2-34. Consideration of risk begins during planning as commanders and staffs assess risk, including ethical risk, for each course of action and propose control measures. They collaborate and integrate input from subordinates, staff, and unified action partners. They determine how to mitigate identified risks. This includes delegating management of certain risks to subordinate commanders who in turn develop appropriate mitigation measures. Commanders then allocate the resources they deem appropriate to mitigate risks.

2-35. Commanders must continuously assess how risk may be accumulating over time as operations progress, both at their own echelon as well as for their subordinates. Changes in the nature of an operation, the number and types of tasks assigned, available combat power, or changes in the threat may all change the level of risk subordinates must mitigate and accept. A series of discrete decisions about seemingly unrelated issues can, over time, potentially change the level of risk in ways that are not readily apparent to a commander. However, this cumulative risk may be understood by one or more subordinates directly impacted by changing conditions or new decisions. It is therefore critical that commanders clearly communicate risk concerns to higher and lower echelons to ensure shared understanding and informed decision making.

2-36. Assumptions initially made during planning may change or compound over time, raising the level of risk. Risks that were acceptable in one context and based on one set of assumptions may be untenable when the context of the operation changes. In some instances, the situation may change to the point that a commander needs to take action to adjust the level of risk subordinate commanders are required to take when the perceived benefit no longer outweighs the likely cost. For example, a unit performing in an economy of force role with a particular task organization may be directed to detach additional units to support other efforts to the point where it can no longer effectively accomplish its mission. The higher level commander is unlikely to have sufficient situational awareness to understand precisely when the threshold of acceptable risk for mission accomplishment has been crossed without a continuous dialogue with that subordinate commander. It is as much the responsibility of the subordinate to keep higher echelons informed as it is the responsibility of the higher level commander to seek risk analysis from the subordinate.

2-37. Inculcating risk acceptance goes hand in hand with creating an environment where subordinates are not only encouraged to take risks, but also one where mistakes are tolerated. Commanders realize that subordinates may not accomplish all tasks initially and that errors may occur. Commanders train subordinates to act within the commander's intent in uncertain situations. Commanders give subordinates the latitude to make mistakes and learn.

2-38. During training, commanders might allow subordinates to execute an excessively risky tactical decision, keeping the safety of Soldiers in mind, as a teaching point. They then instruct subordinates afterward on how to determine a more appropriate level of tactical risk. This sort of coaching helps commanders gain trust in their subordinates' judgment and initiative, and it builds subordinates' trust in their commander. During operations, commanders may need to intervene when a subordinate accepts tactical risk that exceeds the benefits expected.

Risk Acceptance: OPERATION HAWTHORN, Dak To, Vietnam

At 0230, 7 June 1966, a battalion of the 24th NVA (North Vietnamese Army) Regiment attacked an artillery firebase manned by elements of 1st Brigade, 101st Airborne Division, beginning the battle of Dak To. While the forces at the firebase defeated this attack, two battalions of the 101st Airborne were lifted in by helicopters to envelop the 24th NVA Regiment in the Dak To area. One battalion, 1/327th, attacked north up Dak Tan Kan valley, while the other, 2/502d, attacked toward the south. The 1/327th encountered the NVA first and fixed them. The 2/502d established a blocking position initially but then began a sweep south to link up with 1/327th.

The 2/502d used its famous "checkerboard" technique in its advance, breaking down into small units, with squad-size patrols searching designated areas into which the battalion had divided its area of operations. This technique covered ground, but the squads were too weak to face stiff opposition. Company commanders had to assess indicators, decide when they indicated the presence of heavy enemy forces, and assemble their companies for action. As C Company advanced on 12 June, its commander, CPT William Carpenter Jr., sensed those indicators and concentrated his company, but it was surrounded and in danger of being overrun by an estimated NVA battalion. As he spoke to his battalion commander, LTC Hank Emerson ("the Gunfighter"), the sounds of the screaming, charging enemy could be heard over the radio. CPT Carpenter reportedly called for an air strike "right on top of us." The only air support available was armed with napalm; when it hit, it broke the enemy attack and saved the company. A day later, another company linked up with C Company, and they continued the mission. The battle of Dak To was a staggering defeat for the NVA. CPT Carpenter's action can be considered a necessary risk. The survival of his force was at stake. The NVA would have destroyed C Company before another company could relieve it.

CPT Carpenter later stated privately that he realized the survival of his company was at stake, but that he did not actually call the air strike directly in on his position. Instead, he told the forward air controller to use the smoke marking his company's position as the aiming point for the air strike. He knew that using conventional air strike techniques and safe distances would not defeat the enemy. He also reasoned that the napalm would "splash" forward of his position, causing more enemy than friendly casualties. The air strike did just that. Thus, CPT Carpenter exercised judgment based on experience. CPT Carpenter believed he was taking a high risk from the standpoint of troop safety. But he accepted that risk, made a decision, and acted. His actions saved his company and contributed to a major NVA defeat. CPT Carpenter and his first sergeant, 1SG Walter Sabaulaski, received the Distinguished Service Cross for their heroism.

Delegate Authority

> *If it is necessary for a commander to interfere constantly with a subordinate, one or the other should be relieved.*
>
> Field Marshal Richard Michael Carver

2-39. Commanders use judgment to determine how much authority to delegate to subordinates and how much they are able to decentralize execution. Commanders delegate authority and set the level of their personal oversight of delegated tasks based on their assessment of the skill and experience of their subordinates. When delegating authority to subordinates, commanders do everything in their power to set conditions for their success. Commanders allocate sufficient resources to their subordinates so their subordinates can accomplish their missions. Resources include people, units, services, supplies, equipment, networks, information, and time. Commanders allocate resources through task organization and established priorities of support.

2-40. Under the mission command approach, delegated authority is proportional to the extent of commanders' trust in the abilities of their subordinates. Commanders delegate authority and set the level of their personal involvement in delegated tasks based on their assessment of the competence and experience of their subordinates. Ideally, once commanders delegate authority, they supervise to the minimum level required to ensure subordinates' and mission success.

2-41. Commanders delegate authority verbally, in writing through plans, orders, or standard operating procedures, or by both methods. Examples of delegated authority are authority over an area of expertise or technical specialty, a geographic area, or specific kinds of actions. Commanders may limit delegated authority in time, or they may use an enduring approach. Commanders should ensure members of the command, especially the staff and subordinate commanders, understand to whom and what authorities have been delegated. Delegation not only applies to subordinate commanders but also to members of the staff.

Prioritize Resources

2-42. Commanders allocate resources to accomplish the mission. Allocating resources requires judgment because resources can be limited. Considerations for prioritizing resources include how to—

- Effectively accomplish the mission.
- Protect the lives of Soldiers.
- Apply the principles of mass and economy of force.
- Posture their force for subsequent operations.

2-43. The primary consideration for allocating resources is how their use contributes to effective mission accomplishment. Commanders do not determine how to accomplish a mission based on conserving resources or giving all subordinates an equal share; they allocate resources efficiently to ensure effectiveness. The objective—to accomplish the mission—guides every element of operations. A plan that does not accomplish the mission, regardless of how well it conserves resources, is not effective.

2-44. The next priority is to protect the lives of Soldiers. Commanders determine how to protect the lives of Soldiers before considering how to conserve material resources. They use material resources generously to save lives. If there are different but equally effective ways to accomplish the mission, a commander considers ways which use fewer resources.

2-45. The third aspect of resource allocation is based on two of the principles of war—mass and economy of force. The principle of mass means that commanders always weight the main effort with the greatest possible combat power to overwhelm an enemy force and ensure mission accomplishment. Economy of force refers to allocating the minimum essential combat power to all supporting efforts. Supporting efforts typically receive fewer resources than the main effort. Commanders must accept risk in supporting efforts in order to weight the main effort.

2-46. Commanders determine the amount of combat power essential to each task and allocate sufficient resources to accomplish it. When allocating resources, commanders consider the cost to the force and the effects of the current operation on the ability to execute follow-on operations. If subordinates believe they have not received adequate resources, or believe accomplishing their mission would produce an unacceptable cost to the force, they inform their commander. The commander then decides whether to accept risk, allocate more resources, or change the plan.

2-47. The fourth aspect of applying judgment to resource allocation concerns posturing the force for subsequent operations. Commanders balance immediate mission accomplishment with resource requirements for subsequent operations. Commanders accomplish their missions at the least cost to the force, so they do not impair its ability to conduct follow-on operations. They visualize short-term and long-term effects of their resource use and determine priorities. At lower echelons, commanders focus more on the immediate operation—the short term. At progressively higher echelons, commanders give more consideration to long-term operations.

Direct the Staff

2-48. Commanders rely on and expect initiative from the staff as much as from subordinate commanders. Delegating authority allows commanders to use their time for the more creative aspects of command (the art). Commanders delegate authority and set the level of their personal involvement in staff activities based on their assessment of the skill and experience of their subordinates. This assessment requires skilled judgment.

2-49. Within their headquarters, commanders exercise their judgment to determine when to intervene and participate personally in staff operations, as opposed to letting their staffs operate on their own based on guidance. Commanders cannot do everything themselves or make every decision; such participation does not give staffs the experience mission command requires. However, commanders cannot simply approve staff products produced without their input. Commanders participate in staff work where it is necessary to guide their staffs. They use their situational understanding and commander's visualization to provide guidance from which their staffs produce plans and orders. In deciding when and where to interact with subordinates, the key is for commanders to determine where they can best use their limited time to greatest effect—where their personal intervention will pay the greatest dividend.

LEADERSHIP

> *As each man's strengths gives out, as it no longer responds to his will, the inertia of the whole gradually comes to rest on the commander's will alone. The ardor of his spirit must rekindle the flame of purpose in all others; his inward fire must revive their hope.*
>
> Carl von Clausewitz

2-50. *Leadership* is the activity of influencing people by providing purpose, direction, and motivation to accomplish the mission and improve the organization (ADP 6-22). Leadership involves taking responsibility for decisions, being loyal to subordinates, inspiring and directing assigned forces and resources toward a purposeful end, establishing a team climate that engenders success, demonstrating moral and physical courage in the face of adversity, and providing the vision that both focuses and anticipates the future course of events. Leadership requires the attributes and competencies that describe what leaders are expected to be. (See ADP 6-22 for more information on Army leadership.)

2-51. Professional competence, personality, and the will of strong commanders represent a significant part of any unit's combat power. While leadership requirements differ with unit size and type, all leaders must demonstrate character and ethical standards. Leaders must know and understand their subordinates. They must act with courage and conviction in battle. Leaders build trust and teamwork. During operations they know where to be to make decisions or to influence the action by their personal presence.

2-52. Commanders recognize that military operations take a toll on the moral, physical, and mental stamina of the people making up their formations. They understand that experience, interpersonal relationships, and the environment influence the people conducting operations. Leaders account for these factors when motivating people to accomplish tasks in the face of danger and hardship. Setting a good personal example is critical to effective leadership.

2-53. Commanders are both leaders and followers. Being a responsible subordinate is part of being a good leader. Responsible subordinates support the chain of command and ensure their command supports the larger organization and its purpose. Successful commanders recognize the responsibilities they and their subordinates have to the next higher echelon and the larger formation overall.

2-54. Commanders know the status of their forces. Command sergeants major, first sergeants, and platoon sergeants play vital roles in providing commanders awareness about the morale and physical condition of their Soldiers. Commanders need to know when circumstances may prevent friendly forces from performing to their full potential. For example, a subordinate unit may have recently received inexperienced replacements, may have lost cohesion due to leader casualties, or may be extremely fatigued due to an extended period of operations.

2-55. Commanders press the fight tenaciously and aggressively. They accept risks and push Soldiers to the limits of their endurance for as long as possible to retain the operational initiative. They act aggressively to exploit fleeting opportunities. Effective commanders recognize when to push Soldiers to their limits and when to let them rest to prevent individual and collective collapse. Even the most successful combat actions can render units incapable of further operations.

2-56. Leadership is an important component of mitigating the effects of stress. Stress is an integral part of military service in general and in combat operations in particular. Leaders must learn to mitigate stress for their subordinates and cope with it themselves. (See ATP 6-22.5 for more information on combat and operational stress control). Two aspects of leadership that are unique to command are command presence and the location of the commander.

Command Presence

2-57. Command presence is the influence commanders have on those around them through their personal demeanor, appearance, and conduct. It requires contact with others, both physically and through voice command and control systems. Commanders use their presence to gather and share information and assess operations through personal interaction with subordinates. Establishing command presence makes the commander's knowledge and experience available to subordinates and provides encouragement. It allows commanders to assess intangibles like morale, and provide direct feedback on subordinate performance. Commanders employing the mission command approach ask questions without second-guessing their subordinate's performance unless absolutely necessary. Skilled commanders communicate tactical and technical knowledge that goes beyond plans and procedures. Command presence establishes a background for all plans and procedures so that subordinates can understand how and when to adapt them to achieve the commander's intent. Commanders can establish command presence in a variety of ways, including—

- Being seen and heard.
- Sharing risk and hardship.
- Setting a good personal example.
- Ensuring their commander's intent is widely understood.
- Providing clear face-to-face commander's guidance.
- Backbriefs and rehearsals.

2-58. Directly engaging subordinates and staffs allows commanders to motivate, build trust and confidence, exchange information, and assess the human aspects of operations. It allows commanders to assess the morale and stamina of subordinate units. Commanders use their presence to overcome uncertainty and chaos and maintain the focus of their forces. They communicate in a variety of ways, adjusting their communication style to fit the situation and the audience. They communicate both formally and informally, through questions, discussions, conversations, and other direct or indirect communication. Commanders position themselves where they can best command without losing the ability to respond to changing situations.

Location of the Commander

One of the most valuable qualities of a commander is a flair for putting himself in the right place at the vital time.

Field Marshal Sir William Slim

2-59. Commanders command from their personal location. One of the fundamental dilemmas facing all commanders is where to position themselves during operations. As far as conditions allow, commanders should be forward where they can be seen and heard to best influence the decisive operation or main effort. Commanding forward allows commanders to assess the state of operations face to face with their subordinates and achieve shared understanding. It allows them to gather as much information as possible about actual combat conditions when making decisions. However, commanding forward does not mean taking over a subordinate's responsibilities.

2-60. Commanders consider their position in relation to the forces they command and the mission. Their location can have important consequences, not only for the command but also for executing operations. The command and control system helps commanders position themselves forward without losing access to the information and analysis available from their command posts.

2-61. At the lowest levels, commanders lead by personal example, acquire much information themselves, decide personally, and communicate face to face with those they direct. Typically, they position themselves well forward to directly influence the decisive operation. However, even at these levels, commanders cannot always command their whole unit directly.

2-62. In larger tactical- and operational-level commands, command posts are normally the focus of information flow and planning. However, commanders cannot always visualize the battlefield to direct and synchronize operations from command posts. Commanders often assess the situation up front, face to face with subordinate commanders and their units. Commanders employ their command and control system so they can position themselves wherever they can best command without losing the situational understanding that enables them respond to opportunities and changing circumstances. The command and control system allows commanders to obtain the information they need to assess operations and risks, and make necessary adjustments, from anywhere in an area of operations.

2-63. Commanders realize that they might not always be where the critical action is occurring. This probability reinforces the necessity of training subordinates to operate using the mission command approach. Commanders can then rely on subordinates to restore or exploit the situation without their presence or personal intervention.

2-64. Where a commander is located can bring the leadership element of combat power directly to an operation, especially when that location allows for personal presence and the ability to directly observe events and see things that might not be conveyed by the command and control system. Being physically present can allow a commander to assess a much broader set of indicators of the unit's condition. Commanders gain firsthand appreciation for the human aspects of a situation that can rarely be gained any other way. Equally important, commanders can actually see terrain and weather conditions which might not be clearly explained by maps or whether reports. They can avoid the delays and distortions that occur as information travels down and up the chain of command. Finally, by their presence, commanders direct emphasis to critical spots and focus efforts on them. Some of the factors that influence a commander's location include—

- The need to understand the situation.
- The need to make decisions.
- The need to communicate.
- The need to motivate subordinates.

2-65. The commander's forward presence demonstrates a willingness to share danger and hardship. It also allows commanders to appraise for themselves a subordinate unit's condition, including its leaders' and Soldiers' morale. Forward presence allows commanders to sense the human aspects of conflict, particularly when fear and fatigue reduce effectiveness.

2-66. Commanders cannot let the perceived advantages of improved information technology compromise their obligation to personally lead by example. Face-to-face discussions and forward presence allow commanders to see things that may not be conveyed by the command and control system.

THE ROLE OF COMMANDERS IN OPERATIONS

Commanders must remember that the issuance of an order, or the devising of a plan, is only about five per cent of the responsibility of command. The other ninety-five percent is to insure, by personal observation, or through the interposing of staff officers, that the order is carried out.

General George S. Patton, Jr.

2-67. Commanders are the central figures in command and control. Commanders, assisted by their staffs, integrate numerous processes and activities within their headquarters and across the force as they exercise command and control. Throughout operations, commanders balance their time between leading their staffs through the operations process and providing purpose, direction, and motivation to subordinate commanders and leaders.

2-68. The Army's framework for exercising command and control is the *operations process*—the major command and control activities performed during operations: planning, preparing, executing, and continuously assessing the operation (ADP 5-0). Commanders use the operations process to drive the conceptual and detailed planning necessary to understand, visualize, and describe their operational environment and the operation's end state; make and articulate decisions; and direct, lead, and assess operations as shown in figure 2-2.

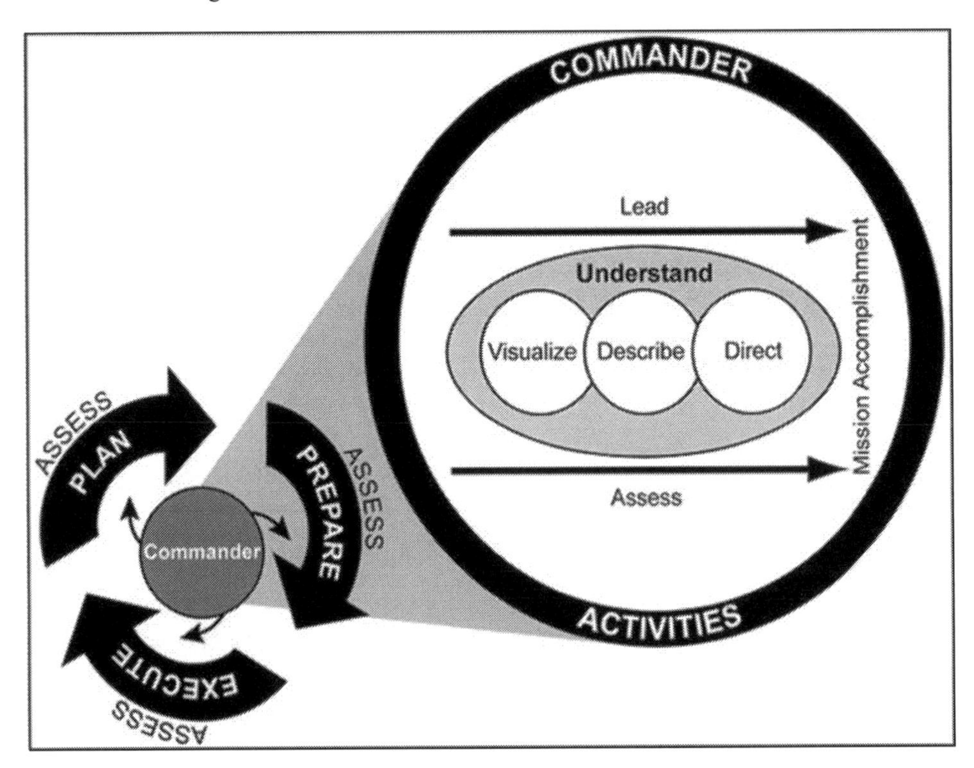

Figure 2-2. The operations process

2-69. The activities of the operations process are not discrete; they overlap and recur as circumstances demand. Planning, preparing, and executing do not always have distinct start and end points. Planning is a continuous activity within the process. While preparing for one operation, or during its execution, units are refining the current plan or planning for future operations. Preparation always overlaps with planning, and it continues throughout execution for some subordinate units. Assessing surrounds and permeates the other three activities as commanders and staffs judge progress toward accomplishing tasks and achieving objectives. Subordinate echelons or units within the same command may be in different stages of the operations process at any given time. (See ADP 5-0 for a detailed discussion of the operations process and the commander's role in planning, preparing, executing, and assessing operations.)

2-70. Commanders, staffs, and subordinate units employ the operations process to integrate and synchronize the warfighting functions across multiple domains and synchronize forces to accomplish missions. This includes integrating numerous processes such as intelligence preparation of the battlefield, the military decision-making process, and the targeting process within the headquarters and with higher echelon, subordinate, supporting, and supported units. The unit's battle rhythm integrates and synchronizes the various processes and activities that occur within the operations process.

2-71. Commanders are the most important participants in the operations process. While staffs perform essential functions that amplify the effectiveness of operations, commanders drive the operations process through understanding, visualizing, describing, directing, leading, and assessing operations as shown in figure 2-3 on page 2-14. Accurate and timely running estimates maintained by staffs assist commanders in understanding the situation and making decisions throughout the operations process.

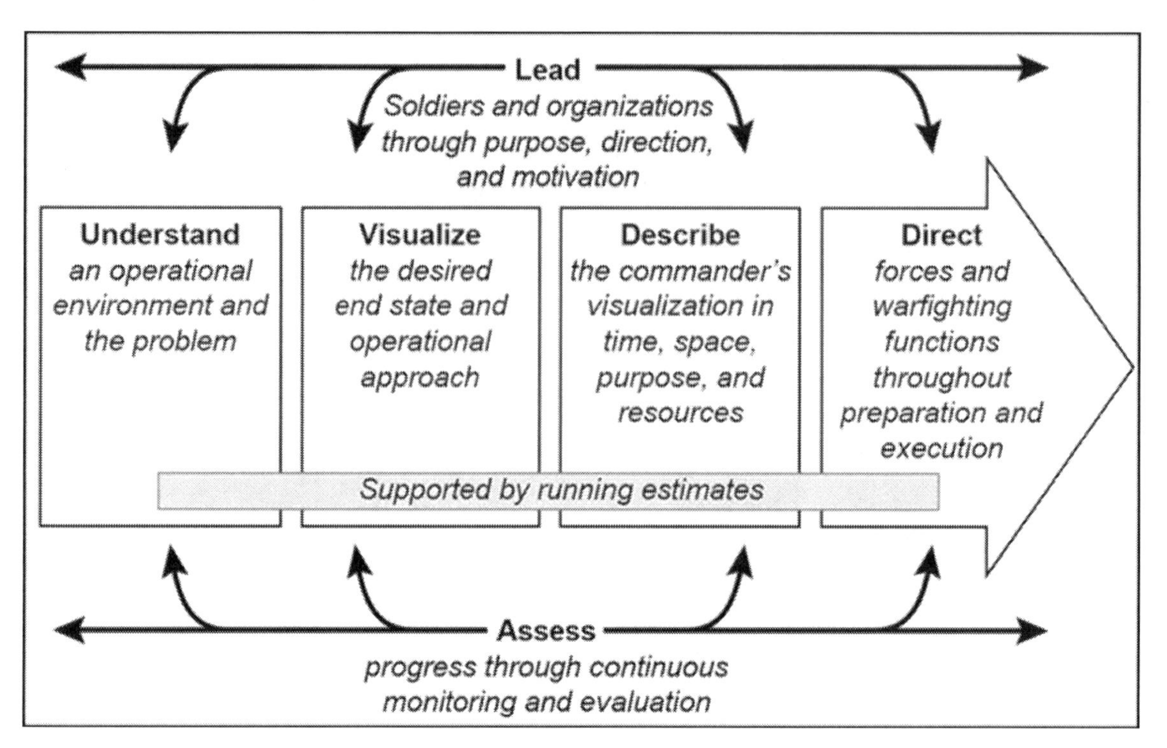

Figure 2-3. The commander's role in the operations process

UNDERSTAND

2-72. An operational environment encompasses physical areas of the air, land, maritime, space, and cyberspace domains as well as the information environment, the electromagnetic spectrum, and other factors. Understanding an operational environment and associated problems is fundamental to establishing a situation's context and visualizing operations. The interrelationship of the air, land, maritime, space, and cyberspace domains and the information environment requires a cross-domain understanding of an operational environment. While understanding the land domain is essential, commanders consider the influence of other domains and the information environment on land operations. They also consider how land power can influence operations in the other domains. For example, commanders consider how friendly and enemy air and missile defense capabilities influence operations in the air domain. Included within these areas are the enemy, friendly, and neutral actors who are relevant to a specific operation.

2-73. Part of understanding an operational environment includes identifying and understanding problems. In the context of operations, an operational problem is a discrepancy between the current state of an operational environment and the desired end state. An operational problem includes those issues that impede commanders from accomplishing missions, achieving objectives, and attaining the desired end state.

2-74. Commanders collaborate with their staffs, other commanders, and unified action partners to build a shared understanding of their operational environment and associated problems. Planning, intelligence preparation of the battlefield, and running estimates help commanders develop an initial understanding of their operational environment. Commanders direct reconnaissance and develop the situation to improve their understanding. Commanders circulate within the area of operations as often as possible, collaborating with subordinate commanders and speaking with Soldiers. Using personal observations and inputs from others (including running estimates from their staffs), commanders improve their understanding of their operational environment throughout the operations process. Ideally, perfect understanding should be the basis for decisions. However, commanders realize that uncertainty and time preclude achieving perfect understanding before deciding and acting.

VISUALIZE

2-75. As commanders begin to understand their operational environment, they start visualizing a desired end state and potential solutions to solve or manage identified problems. Collectively, this is known as ***commander's visualization*—the mental process of developing situational understanding, determining a desired end state, and envisioning an operational approach by which the force will achieve that end state**. Figure 2-4 depicts activities associated with developing the commander's visualization.

2-76. In building their visualization, commanders first seek to understand those conditions that represent the current situation. Next, commanders envision a set of desired future conditions that represents the operation's end state. Commanders complete their visualization by conceptualizing an *operational approach*—a broad description of the mission, operational concepts, tasks, and actions required to accomplish the mission (JP 5-0).

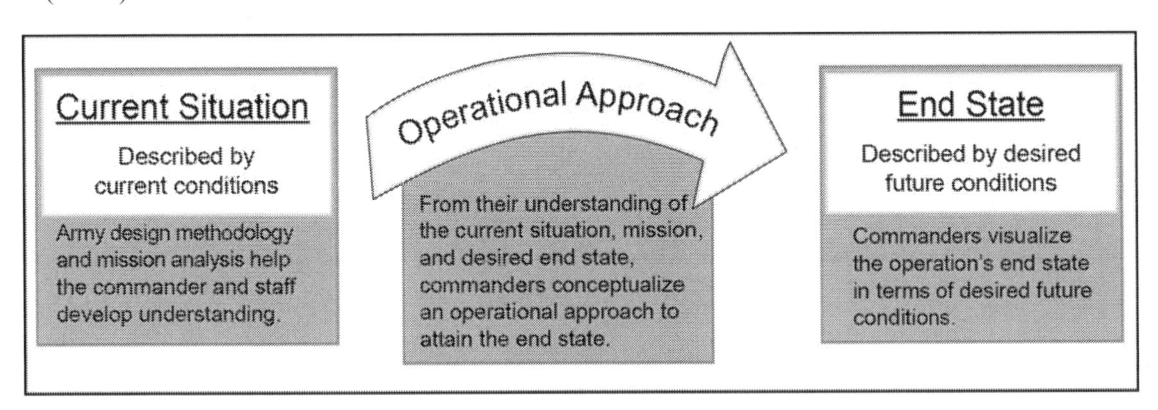

Figure 2-4. Commander's visualization

DESCRIBE

2-77. Commanders describe their visualization to their staffs and subordinate commanders to facilitate shared understanding and purpose throughout the force. During planning, commanders ensure subordinate commanders understand their visualization well enough to begin course of action development. During execution, commanders describe modifications to their visualization in updated planning guidance and directives resulting in fragmentary orders that adjust the original operation order. Commanders describe their visualization in doctrinal terms, refining and clarifying it as circumstances require. Commanders describe their visualization in terms of—

- Commander's intent.
- Planning guidance, including an operational approach.
- Commander's critical information requirements (CCIRs).
- Essential elements of friendly information.

DIRECT

2-78. Commanders direct action to achieve results and lead forces to mission accomplishment. Commanders make decisions and direct action based on their situational understanding maintained by continuous assessment. Throughout the operations process, commanders direct forces by—

- Preparing and approving plans and orders.
- Establishing command and support relationships.
- Assigning and adjusting tasks, control measures, and task organization.
- Positioning units to maximize combat power.
- Positioning key leaders at critical places and times to ensure supervision.
- Allocating resources to exploit opportunities and counter threats.
- Committing the reserve as required.

Lead

2-79. Commanders lead by example and personal presence. Leadership inspires subordinates to accomplish things that they otherwise might not. Where a commander locates within an area of operations is an important consideration for effective mission command. Through leadership, commanders provide purpose, direction, and motivation to subordinate commanders, their staffs, and Soldiers. There is no standard pattern or simple prescription; different commanders lead differently. Commanders balance their time among command posts and staffs, subordinate commanders, forces, and other organizations to make the greatest contribution to success. (See ADP 6-22 for a detailed discussion of leadership)

Assess

2-80. Commanders continuously assess the situation to better understand current conditions and determine how an operation is progressing. Continuous assessment helps commanders anticipate and adapt the force to changing circumstances. Commanders incorporate the assessments of their staffs, subordinate commanders, and unified action partners into their personal assessment of the situation. Based on their assessment, commanders adjust their visualization and modify plans to adapt the force to changing circumstances.

2-81. A commander's focus on understanding, visualizing, describing, directing, leading, or assessing throughout operations varies during different operations process activities. For example, during planning commanders focus more on understanding, visualizing, and describing. During execution, commanders often focus more on directing, leading, and assessing—while improving their understanding and modifying their visualization as needed. (See ADP 5-0 for a detailed discussion on assessing operations.)

GUIDES TO EFFECTIVE COMMAND

There will be neither time nor opportunity to do more than prescribe the several tasks of the several subordinates.... [I]f they are reluctant (afraid) to act because they are accustomed to detailed orders... if they are not habituated to think, to judge, to decide and to act for themselves ... we shall be in sorry case when the time of "active operations" arrives.

Fleet Admiral Ernest J. King

2-82. The guides to effective command help commanders fulfill the fundamental responsibilities of command. A commander's use of these guides must fit the situation, the commander's personality, and the capability and understanding of subordinates. Command cannot be scripted. These guides apply at all levels of command. Mission command provides a common baseline for command during operations and garrison activities. These guides aid commanders in effectively exercising command and inculcating mission command:

- Create a positive command climate.
- Ensure unity of effort.
- Train subordinates on command and control and the application of mission command.
- Make timely and effective decisions and act.

Create a Positive Command Climate

Morale is a state of mind. It is that intangible force which will move a whole group of men to give their last ounce to achieve something, without counting the cost to themselves; that makes them feel they are part of something greater than themselves.

Field Marshall Sir William Slim

2-83. Commanders create their organization's tone—the characteristic atmosphere in which people work. This is known as the command climate. It is directly attributable to the leader's values, skills, and actions. A positive climate facilitates team building, encourages initiative, and fosters collaboration, mutual trust, and shared understanding. Commanders shape the climate of their organization, no matter what the size.

2-84. Successful commanders recognize that all subordinates contribute to mission accomplishment. They establish clear and realistic goals and communicate their goals openly. Commanders establish and maintain open, candid communication between subordinate leaders. They encourage subordinates to bring creative and innovative ideas to the forefront. They also seek feedback from subordinates. The result is a command climate that encourages initiative.

2-85. A positive command climate instills a sense of trust within units. It facilitates a strong sense of discipline, comradeship, self-respect, and morale. It helps Soldiers develop a desire to do their fair share and to help in the event of need. In turn, Soldiers know their leaders will guard them from unnecessary risk.

2-86. In a positive command climate, the expectation is that everyone lives by and upholds the moral principles of the Army Ethic. The Army Ethic must be espoused, supported, practiced, and respected. Mission command depends on a command climate that encourages subordinate commanders at all levels to take the initiative. Commanders create a positive command climate by—

- Accepting subordinates' risk taking and errors.
- Building mutual trust and shared understanding.
- Communicating with subordinates.
- Building teams.

(See AR 600-20, ADP 6-22, and FM 6-22 for more information on creating a positive command climate.)

Accept Subordinates' Risk Taking and Errors

Judgment comes from experience and experience comes from bad judgment.

General Omar N. Bradley

2-87. Exercising initiative requires a command climate that promotes risk taking. Commanders inculcate the willingness to accept risk into their commands through leading by example and accepting subordinates' risk taking. They accept risk during training and operations. They assess the judgment of their subordinates' risk taking, either at the time of decision, if time permits, or during after action reviews.

2-88. Commanders allow subordinates to accept risk. In training, commanders might allow subordinates to execute an excessively risky decision, as long as it does not endanger lives, as a teaching point. This training helps commanders gain trust in their subordinates' judgment and initiative and builds subordinates' trust in their commander.

2-89. Inculcating risk acceptance among subordinates requires that commanders accept risk themselves. Subordinates will not always succeed, particularly when inexperienced. However, with risk acceptance in the command climate, subordinates learn, gaining the experience required to operate on their own. In addition, subordinates learn to trust their commander to give them authority to act, knowing their commander will back their decisions.

2-90. Commanders do not underwrite subordinate mistakes when a subordinate operates outside of the commander's intent or displays poor judgment that endangers life or mission accomplishment. Nor do commanders tolerate a subordinate who repeatedly makes the same mistakes, does not learn, or violates the Army Ethic. Discriminating between which mistakes to underwrite as teaching points and which mistakes are unacceptable in a military leader is part of the art of command.

Build Mutual Trust and Shared Understanding

2-91. Mutual trust and shared understanding are critical to subordinates' exercise of initiative. Mutual trust and shared understanding of the commander's intent frees commanders to move about the battlefield. Commanders know their subordinates understand the desired end state, and subordinates know their commander will support their decisions. Additionally, this command climate allows commanders to operate, knowing subordinates will accurately and promptly report both positive and negative information. Mutual trust and shared understanding are critical to the tempo of large-scale combat operations.

Mutual Trust and Shared Understanding: VII Corps and the Ruhr Encirclement

First Army's VII Corps, under MG J. Lawton Collins, entered action in Europe on 6 June 1944. MG Collins' staff served with him almost uninterruptedly before and through the campaign. This familiarity helped ensure that MG Collins' subordinates would understand and carry through his intent in issuing and executing their own orders. MG Collins' command techniques supported subordinates' exercise of initiative. He discussed his principal decisions, important enemy dispositions, and principal terrain features with major subordinate commanders. If he could not assemble these commanders, he visited them individually as time permitted, with priority given to the commander of the decisive operation. During operations, he visited major subordinate units to obtain information on enemy reactions and major difficulties encountered, again giving priority to units conducting the decisive operation. His general and special staff officers visited other units to report critical matters to the corps chief of staff. Upon returning to headquarters, MG Collins met with his staff to review the day's events and the changes he had directed. After that, his G-3 prepared and distributed a daily operations memorandum confirming MG Collins' oral instructions and adding any other information or instructions developed during the staff meeting. During the European campaign, VII Corps issued only 20 field orders, an average of two per month, to direct operations.

For the Ruhr encirclement, First Army's mission was to break out from its Rhine River bridgehead at Remagen, link up with Third Army in the Hanau-Giessen area, and join Ninth Army of 21st Army Group near Kassel-Paderborn. The attack began on 25 March 1945, with VII Corps attacking and passing through the enemy's main defensive positions. By this time, GA Dwight D. Eisenhower, Supreme Commander Allied Expeditionary Force, had decided to isolate the Ruhr from north and south by encirclement, the junction point being the Kassel-Paderborn area. On 26 March, VII Corps took Altenkirchen and, on 27 March, crossed the Dill River. First Army assigned VII Corps as the decisive operation for the linkup with Ninth Army at Paderborn. MG Collins had only 3d Armored Division and 104th Infantry Division available, and the objective was more than 100 kilometers away. Nevertheless, 3d Armored Division, commanded by MG Maurice Rose, was directed to reach Paderborn in one day, and MG Rose, in turn, assigned his subordinates decisive and shaping operations to accomplish that mission. The decisive operation halted 25 kilometers short of Paderborn at 2200 on 29 March. The next day MG Rose was killed in action. As the Germans strongly defended Paderborn, 3d Armored Division's lead elements were held 10 kilometers from the town. The corps received intelligence of German counterattack forces building around Winterberg, southwest of Paderborn. To counter this, 104th Infantry Division took the road junctions of Hallenberg, Medebach, and Brilon. First Army ordered III and V Corps to shield VII Corps from any attacks from outside the ring.

As the situation developed, MG Collins adapted the corps plan to his situational understanding, while remaining within the framework of the higher echelon commander's intent. By 31 March, German attacks against 104th Infantry Division, increasing German resistance around Paderborn, 3d Armored Division's reorganization necessitated by MG Rose's death, and preparation of a coordinated attack against Paderborn required MG Collins to contact the Ninth Army commander and suggest a change in the linkup point. They agreed on the village of Lippstadt, halfway between Paderborn and the lead elements of 2d Armored Division (the right-flank division of Ninth Army). The linkup was effected on 1 April, closing the Ruhr pocket. MG Collins personally led a task force from 3d Armored Division, overcoming weak resistance in its push west, linking up with elements of 2d Armored Division at 1530 at Lippstadt. Later that day, VII Corps successfully overcame the German defenses at Paderborn. The encirclement trapped Army Group B, including Field Marshal Model, 5th Panzer and 15th Armies, and parts of 1st Parachute Army, along with seven corps, 19 divisions, and antiaircraft and local defense troops—a total of nearly 350,000 soldiers. The reduction of the Ruhr pocket would take another two weeks. (See figure 2-5.)

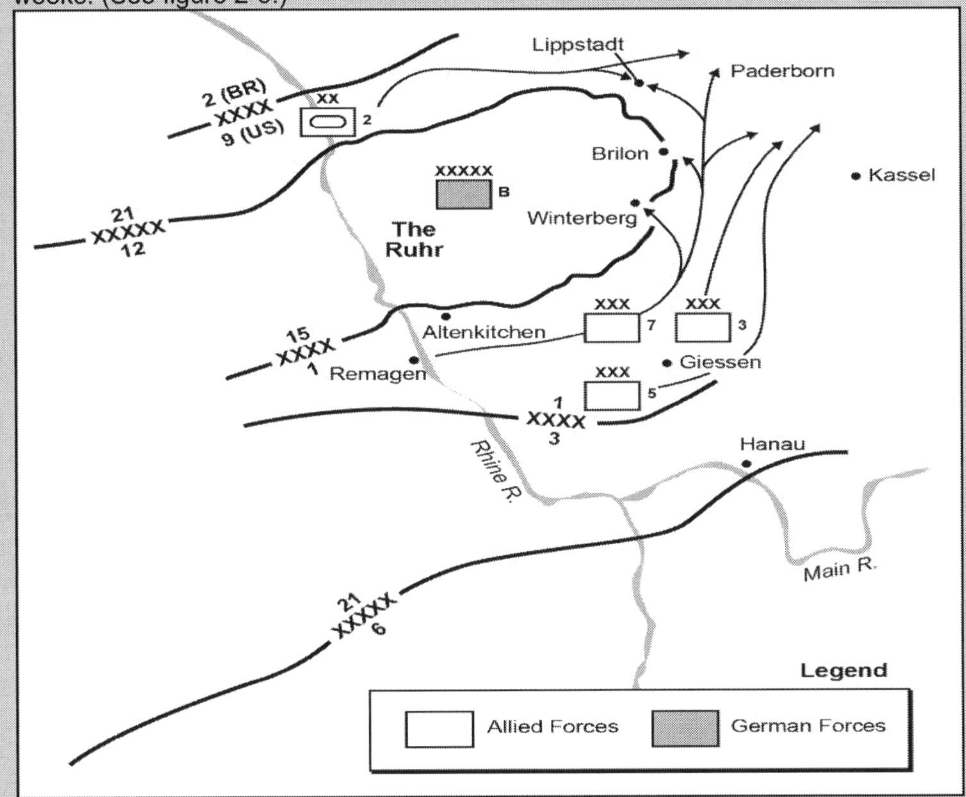

Figure 2-5. Map of Ruhr encirclement

> The Ruhr had been selected as an objective even before the Allies landed in Europe. All major commanders appear to have understood this. However, 12th Army Group only gave the actual orders for the encirclement in late March 1945, when the success of First Army's breakout had become clear. The actual linkup was eventually effected between VII Corps and Ninth Army, principally on MG Collins' understanding of the higher echelon commander's intent and initiative by his subordinates. He practiced a technique similar to mission orders, giving only one or two immediate objectives to each major subordinate command and a distant objective toward which to proceed, without specific instructions. This gave his subordinates freedom to act and exercise initiative, while still providing essential elements needed for coordination among the subunits. Knowing the overall commander's intent enabled commanders on both sides of the encirclement to direct efforts toward its fulfillment. When lack of lateral communications hindered coordination, subordinates took the initiative to accomplish the mission and fulfill the commander's intent as they understood it. At 3d Armored Division, subordinates' understanding of the corps' commander's intent allowed operations to resume the day after Rose was killed. When the original objective, achieving a linkup at Paderborn, could no longer be accomplished, MG Collins proposed an alternative linkup point. Finally, with elements of his corps defending at Winterberg, attacking at Paderborn, and moving to Lippstadt, MG Collins positioned himself with the task force from 3d Armored Division to make the decisive operation that day for his corps, First Army, and 12th Army Group.

Communicate with Subordinates

> *General Meade was an officer of great merit, with drawbacks to his usefulness that were beyond his control.... [He] made it unpleasant at times, even in battle, for those around him to approach him even with information.*
>
> General Ulysses S. Grant

2-92. Communicating with subordinates contributes to the shared understanding fundamental to mission command. Effective commanders take positive steps to encourage, rather than impede, communications among and with their subordinates and staff. Candor and the free exchange of ideas contribute to trust. Commanders make themselves available and accessible for communications and open to new information. They create a climate where collaboration routinely occurs throughout their organization through personal example, coaching, and mentorship.

2-93. Successful commanders invest the time and effort to visit and engage with Soldiers, subordinate leaders, and unified action partners to understand their issues and concerns. Through these interactions, subordinates and partners gain insight into the commander's leadership style and concerns.

Build Teams

2-94. Army organizations rely on effective teams to complete tasks, achieve objectives, and accomplish missions. The ability to build and maintain effective teams throughout military operations is an essential skill for all Army commanders, staffs, and leaders.

2-95. *Army team building* is a continuous process of enabling a group of people to reach their goals and improve effectiveness through leadership and various exercises, activities and techniques (FM 6-22). The goal of Army team building is to improve the quality of the team and how it works together to accomplish the mission. Using doctrinal terms and symbols is one method of fostering teamwork. Often, the only basis for trust and teamwork in situations that require rapid task organization is a common language and approach to operations. Training and rehearsals also provide opportunities to foster teamwork. Teambuilding is essential to achieving the effective teams required of mission command. (See ATP 6-22.6 for more information on teams and teamwork.)

2-96. Building an effective team is challenging, but the positive benefits of teamwork in an effective team are well worth the effort and time it takes. These benefits enhance the performance of the team, improve the skills of the individual team members, and build important relationships with unified action partners.

2-97. Commanders must foster teamwork among task-organized units. Commanders initiate team building, both inside and outside their organizations, as early as possible, and they maintain it throughout operations. Commanders must trust and earn the trust of their unified action partners and key leaders within the operational area. Overcoming differences in cultures, mandates, and organizational capabilities is key to building mutual trust with unified action partners and key leaders.

2-98. Commanders use interpersonal relationships to build teams within their own organizations and with unified action partners. Uniting all the diverse capabilities necessary to achieve success in operations requires collaborative and cooperative efforts that focus those capabilities toward a common goal. Where military forces typically demand unity of command, a challenge for building teams with unified action partners is to forge unity of effort—coordination and cooperation toward common objectives. (See JP 3-08 for more information on teambuilding with unified action partners.)

Ensure Unity of Effort

2-99. *Unity of effort* is the coordination and cooperation toward common objectives, even if the participants are not necessarily part of the same command or organization, which is the product of successful unified action (JP 1). Establishing a culture of collaboration provides and enhances unity of effort. The commander's intent provides the unifying idea that allows decentralized execution within an overarching framework.

2-100. Unity of command is one of the principles of war and the preferred method for achieving unity of effort. Commanders always adhere to unity of command when task-organizing Army forces. Under unity of command, every mission falls within the authority and responsibility of a single, responsible commander. Unity of command requires that two commanders may not exercise the same command relationship over the same force at any one time.

2-101. Unity of command may not be possible in some operations that include unified action partners. When unity of command is not possible, commanders must achieve unity of effort through cooperation and coordination to build trust among all elements of the force—even if they are not part of the same command structure.

2-102. The commander's intent provides guidance within which subordinates are expected to exercise initiative to accomplish overall goals. Understanding the commander's intent two echelons up further enhances unity of effort while providing the basis for decentralized decision making and execution. Subordinates who understand the commander's intent are more likely to exercise disciplined initiative in unexpected situations. Under mission command, subordinates have an absolute responsibility to fulfill the commander's intent.

Train Subordinates in Command and Control and the Application of Mission Command

2-103. Commanders develop a basic level of control within their organizations when they create a culture that embraces mission command at every level. The time spent inculcating mission command into training, education, and problem solving prior to operations saves time and simplifies command and control during operations. Commanders cannot expect subordinates to respond effectively to a mission command approach once operations commence if they have not developed subordinates comfortable in its use beforehand.

2-104. Leaders have an obligation to ensure that their subordinates are capable of performing their assigned tasks to Army standards under a variety of circumstances. Leaders generally provide more direction or guidance and control until they are satisfied that subordinates understand tasks, conditions, and standards and can operate within the commander's intent. Increased confidence in the ability of subordinates generally leads to more latitude in the way that subordinates are given to complete their assigned missions, since commanders can trust that the subordinates understand the purpose of what they are being told to do.

2-105. The ability to provide general guidance oriented on the purpose of a mission saves time during execution for both commanders and subordinates and maximizes flexibility should conditions change or communication become intermittent. It also minimizes the chances that subordinates will waste resources on tasks no longer relevant to the purpose of a particular operation. The effort put into developing subordinate leaders and their teams saves critical time in combat and allows commanders to assume more tactical risk when the situation is unclear and communication is intermittent.

2-106. Effective mission command requires well-developed subordinates capable of decentralized execution of missions and tasks. Training must create common, repetitive, shared experiences that build trust and allow commands to acquire competence in shared understanding. Trained teams are able to communicate explicitly and implicitly, conduct decentralized operations, and achieve unity of effort in uncertain situations.

2-107. Noncommissioned officers are key enablers of mission command, and they must be trained in the mission command principles to effectively support their commander and lead their Soldiers. Noncommissioned officers are required to exercise disciplined initiative to make decisions and take actions to further their commander's intent. They must actively work to understand the commander's intent two levels up and relay that intent to their Soldiers. They train to develop mutual trust and shared understanding with their commanders and their Soldiers.

2-108. Noncommissioned officers enforce standards and discipline and develop their subordinates as they build teams. They are trained to operate under mission orders and decide for themselves how best to achieve their commander's intent. With information available to all levels of command and increasing dispersion on the battlefield, noncommissioned officers must be comfortable in exercising initiative to make decisions and act.

2-109. As part of training subordinates in command and control and mission command, commanders must prepare their subordinates for positions of increased responsibility. To this end, they promote leader qualities and assess subordinates' potential for future appointments to command and staff positions.

Promote Leader Qualities

2-110. Commanders promote leader qualities by developing them in themselves and in their subordinates at least two echelons down. These qualities are described in the Army leadership requirements model. This model describes the attributes and competencies required of Army leaders. But qualities alone do not make successful commanders. Successful commanders develop a balance among those qualities. The fact that an officer has been appointed a commander does not automatically endow that commander with these qualities. Rather, all officers develop them to prepare for command. In general, the higher the echelon of command, the wider the scope of qualities required. In addition, the emphasis on and among the qualities changes with the echelon of command. (See ADP 6-22 for more information on the Army leadership requirements model.)

2-111. All commanders emphasize the Warrior Ethos. The Warrior Ethos is a set of principles by which every Soldier lives, and it states—

- I will always place the mission first.
- I will never accept defeat.
- I will never quit.
- I will never leave a fallen comrade behind.

2-112. The Warrior Ethos is perishable, so commanders continually affirm, develop, and sustain it. Developing it demands inculcating self-discipline in the commander, subordinates, and the command. It requires tough, realistic training that develops the resiliency needed to endure extremes of weather, physical exertion, and lack of sleep and food. Commanders develop the will, determination, and the confidence that they, their subordinates, and their formations will accomplish all missions regardless of conditions.

2-113. Training and education can develop much of the knowledge and many of the skills commanders require. In particular, training devices, simulations, and exercises can enhance clarity of thought and judgment, including decision making. Developing leader qualities and practicing leadership skills is necessary for subordinates to decide and act effectively during operations.

Assess Subordinates

No man is more valiant than Yessoutai; no one has rarer gifts. But, as the longest marches do not tire him, as he feels neither hunger nor thirst, he believes that his officers and soldiers do not suffer from such things. That is why he is not fitted for high command.

Genghis Khan

2-114. Once appointed, commanders assume the role of coach and mentor to their subordinates. They study the personalities and characteristics of their subordinate commanders. Some need significant direction; others work best with little or no guidance. Some tire easily and require encouragement and moral support. Others, perhaps uninspired in peace, excel in conflict and war. Matching talent to tasks is an important function of command. Commanders judge Soldiers so they can appoint the right subordinates to the right positions at the right times. Assessing individuals and handling them to the best effect applies to staffs as well as subordinate commanders. Commanders also assess subordinates by giving them experiences and opportunities to grow through assignments that challenge their abilities. Recognizing subordinates' strengths and limits is vital to effectively exercising command.

2-115. One of a commander's most important duties is evaluating subordinates to identify talent—potential future candidates for senior appointments to command and staff positions. To assess the command qualities of subordinates objectively, commanders place individuals in circumstances where they must make decisions and live with the consequences. In these situations, subordinates must know their commander has enough confidence in them to permit honest mistakes. Training gives commanders opportunities to assess subordinates on the qualities commanders should possess. In particular, assessing subordinates should confirm whether they exhibit the necessary balance of intelligence, professionalism, and common sense required to carry the added responsibilities that go with promotion.

2-116. An important aspect of assessing subordinates is determining the extent to which they are both willing and able to apply the mission command approach to the command and control of their units. Since commanders evaluate two echelons down during training and leader development, and observe subordinate leader behavior one and two echelons down during training and operations, under both garrison and field conditions, they have multiple opportunities for assessing internalization of the mission command approach. (See FM 6-22 for further discussion on assessing subordinates.)

Make Timely Decisions and Act

I have found again and again that in encounter actions, the day goes to the side that is the first to plaster its opponent with fire. The man who lies low and awaits developments usually comes off second best.

Field Marshal Erwin Rommel

2-117. Timely decisions and actions are essential for effective command and control. Commanders who demonstrate the agility to consistently make appropriate decisions faster than their opponents have a significant advantage. By the time the slower commander decides and acts, the faster one has already changed the situation, rendering the slower commander's actions irrelevant. With such an advantage, the faster commander can dictate the tempo and maintain the operational initiative.

2-118. A mission command approach makes it easier for commanders to make timely decisions that exploit opportunities because they spend less time focused on subordinates' tasks. Effective commanders—

- Take the enemy situation, capabilities, and reaction times into account when making decisions.
- Consider the impact of their decisions—the cause and effect.
- Make decisions quickly—even with incomplete information.
- Adopt a satisfactory course of action with acceptable risk as quickly as possible.
- Delegate decision making authority to the lowest echelon possible to obtain faster decisions during operations.
- Support decentralized execution by maintaining shared understanding with subordinates and frequently with adjacent commanders.

2-119. Commanders change and combine intuitive and analytical decision making techniques as the situation requires. Because uncertainty and the tempo of large-scale combat operations drive most decisions, commanders emphasize intuitive decision making and develop their subordinates accordingly. However, when time is available and depending on the operational context of a situation, commanders and staffs use the military decision-making process or Army design methodology during planning.

2-120. Commanders can alter planning to fit time-constrained circumstances. In time-constrained conditions, commanders assess the situation, update their commander's visualization, and direct their staffs to perform those activities needed to support the required decisions. Streamlined processes permit commanders and staffs to shorten the time needed to issue orders when the situation changes. To an outsider, it may appear that experienced commanders and staffs omit key steps. In reality, they use existing products or perform steps mentally. Commanders ensure their staffs are trained on all Army decision-making methodologies. (See ADP 5-0 for more information on the Army decision-making process.)

2-121. Commanders and staffs constantly assess where an operation is in relation to the end state and make adjustments to accomplish the mission and posture the force for future operations. The commander's visualization and the staffs' running estimates are the primary assessment tools. Keeping running estimates current is essential to ensuring commanders are aware of feasible options. Staffs continuously replace outdated facts and assumptions in their running estimate with new information. They perform analysis and form new, or revise existing, conclusions and recommendations. The commander's visualization identifies decisions commanders expect to make and focuses their staffs' running estimates. Up-to-date running estimates provide the recommendations commanders need to make timely decisions during execution. (See FM 6-0 for more information on running estimates.)

CONCLUSION

Good morale and a sense of unity in a command cannot be improvised; they must be thoroughly planned and systematically promoted. They are born of just and fair treatment, a constant concern for the soldier's welfare, thorough training in basic duties, comradeship among men and pride in self, organization, and country. The establishment and maintenance of good morale are incumbent upon every command and are marks of good leadership.

FM 100-5, *Operations* (1941)

2-122. The role of commanders is to direct and lead from the beginning of planning throughout execution, and continually assess and adjust operations to achieve their intent. Commanders drive the operations process. They understand, visualize, describe, direct, lead, and assess operations in complex, dynamic environments. Throughout operations, commanders, subordinate commanders, staffs, and unified action partners collaborate actively, sharing and questioning information, perceptions, and ideas to better understand situations and make decisions. Commanders encourage disciplined initiative through mission orders and a climate of mutual trust and shared understanding. Guided by their experience, knowledge, education, intelligence, and intuition, commanders apply leadership to translate decisions into action. Commanders synchronize forces and capabilities in time, space, and purpose to accomplish missions.

2-123. Ultimately, command reflects everything a commander understands about the nature of war, warfighting doctrine, training, leadership, organizations, materiel, and soldiers. It is how commanders organize their forces, structure operations, and direct the synchronized effects of organic and allocated assets toward their visualized end state. Command is built on training and shared understanding by all Soldiers within a command about how it operates. It is the expression of the commander's professional competence and leadership style, and the translation of the commander's vision to the command. However, command alone is not sufficient to translate that vision and to assure mission accomplishment; control, the subject of chapter 3, is also necessary.

Chapter 3
Control

The test of control is the ability of a leader to obtain the desired result from his command.
Infantry in Battle, 1939

This chapter begins with a discussion of the nature of control and its elements. It next addresses the various types of control measures. The chapter concludes with a discussion of guides for effective control.

NATURE OF CONTROL

Everything in war is very simple, but the simplest thing is difficult. The difficulties accumulate and end by producing a kind of friction that is inconceivable unless one has experienced war....Friction...makes the apparently easy so difficult.
Carl von Clausewitz

3-1. Within command and control, control is the regulation of forces and warfighting functions to accomplish the mission in accordance with the commander's intent. Commanders exercise control to direct and adjust operations as conditions dictate. Unlike aspects of command, which remain relatively similar among echelons, control functions increase in complexity at each higher echelon. Control extends over the entire force and may include control of the airspace over an area of operations below the coordinating altitude. (See FM 3-52 for more discussion of airspace control).

3-2. Control allows commanders to monitor and receive feedback regarding the situation during operations. Based on this information, commanders can modify their visualization and direct changes to an operation as necessary. In the broadest terms, control helps commanders answer two questions:

- What is the actual situation compared with the desired end state?
- Are adjustments to the plan necessary to reconcile the situation with the desired end state?

3-3. Control, as contrasted with command, is more science than art. As such, it relies on objectivity, facts, empirical methods, and analysis. The science of control supports the art of command. Commanders and staffs use the science of control to understand the physical and procedural constraints under which units operate. Units are bound by such factors as movement rates, fuel consumption, weapons effects, rules of engagement, and other legal considerations. Commanders and staffs strive to understand aspects of operations they can analyze and measure, such as the physical capabilities and limitations of friendly and enemy organizations.

3-4. Control requires a realistic appreciation of time and distance factors, including the time required to initiate, complete and assess directed actions. There is art in anticipating likely points of friction and factors beyond subordinate control that invariably lead to delays during execution. The higher the echelon and the larger a formation, the longer it takes to complete assigned tasks, which in turn requires an earlier decision to achieve a desired effect or end state. The planning necessary to facilitate adequate control should always incorporate time tolerances that account for the friction inherent in operations.

3-5. Commanders, aided by staffs, use control to regulate forces and the functions of subordinate and supporting units. Staffs give commanders their greatest support in providing control. However, for control to be effective, commanders must actively participate in exercising it.

3-6. One of mission command's strengths is that it provides a measure of self-regulation during the conduct of operations. Under mission command, control tends to be decentralized and flexible whenever possible in the context of operations. Orders and plans rely on subordinates' abilities to coordinate among themselves to create shared understanding and synchronize operations. By delegating decision-making authority to facilitate decentralized execution, mission command increases tempo by improving a subordinate's ability to act in rapidly changing situations. As a result, the mission command approach to command and control is inherently less vulnerable to disrupted communications.

Levels of Control and German Auftragstaktik

German tactical successes throughout the early years of WWII are often touted as a prime example of an unfettered mission command style with few considerations of control. In practice this was not actually true. The German emphasis on decentralization and initiative came from their tradition of Auftragstaktik, or mission-oriented tactics. It was a military culture that relied on subordinate commanders and junior leaders to recognize the overall intent of a mission, and take necessary actions to ensure the mission intent was met even if actions appeared to countermand prior guidance and orders. The German campaign against France in May of 1940 is a case study of positive and negative applications of the German army's approach to command and control. In 46 days, the German army routed Allied forces and established control of the whole of Western Europe. This happened in part because of the weaknesses of their adversaries, particularly the French.

French doctrine emphasized control by senior level commanders to enable "methodical battle." It required carefully planned and synchronized employment of fires and maneuver forces, and essentially relied upon a centralized, deliberate approach at every echelon. Such an approach assumed excellent communications, good situational awareness, and a similarly deliberate approach by the enemy. Methodical battle was ill-suited for situations requiring rapid decision making and initiative at all echelons, situations that occurred repeatedly between 10 and 20 May 1940.

While Auftragstaktik ensured tactical commanders had the flexibility to adjust their plans as necessary, there were negative repercussions associated with the undisciplined use of initiative beyond the constraints of the commander's intent. Because some subordinate German commanders felt unduly restricted by their assigned routes, they failed to direct the proper march discipline for their difficult movement through the Ardennes. This disregard for movement plans threatened to disrupt complex march tables essential to moving large forces through the restrictive terrain of the Ardennes Forest. There was little room for initiative under such conditions, unless it was exercised in support of the plan as devised. Any deviation from the plan would have second and third order effects throughout the column. On the morning of the second day of the campaign, von Kleist sent out a message to his subordinate commanders that said the problems encountered during the move through Luxembourg were caused primarily by independent decisions being made by lower level leaders, and if conditions worsened offenders would be punished with the death penalty.

The German XIX Corps commander, GEN Heinz Guderian, repeatedly clashed with his superior, GEN Von Kleist, commander of the Panzer Gruppe, with regard to the movements and orientation of his corps. He achieved significant positive results during the first week of the operations, even as he argued with his commander and disobeyed or ignored multiple directives intended to reduce risk to the overall operation. At the end of the first week of fighting, GEN Guderian offered his resignation during a disagreement about the necessity for a short tactical pause after crossing the Meuse. To his surprise, GEN von Kleist accepted it, an indicator that even in the German Army challenging the constraints of a higher commander's intent had consequences. If GEN Guderian had not been so difficult during the initial phases of the operation where control was critical to its overall success, he may have enjoyed greater support at the critical point where initiative and risk taking would contribute more to achieve operational and strategic level success.

Effective application of the mission command approach requires judgment in establishing the degree of control required in a particular situation. Once German forces were engaged in close combat with French forces along the Meuse River, the focus was to cross as rapidly as possible, which meant that subordinate commanders were expected to demonstrate initiative continuously and lead from the front. Circumstances were much different than they were on the approach march. Balance between obedience and initiative must be struck in every unique context, and effectively achieving that balance is central to mission command.

ELEMENTS OF CONTROL

3-7. Commanders use control to direct and coordinate the actions of subordinate forces to meet their intent. They communicate information and receive feedback from subordinates to achieve greater shared understanding of the situation. This allows commanders to update their visualization with respect to the current situation, the end state or their operational approach, and adjust operations to reflect those changes. The elements of control are—

- Direction.
- Feedback.
- Information.
- Communication.

3-8. Command and control, specifically control, is not a one-way activity in which commanders direct while subordinates comply with orders. In application, command and control is multidirectional, with feedback and influence from sources below, above, laterally, as well as outside the chain of command, as shown in figure 3-1 on page 3-4. It includes the reciprocal flow of information between commanders, staffs, subordinate forces and other external organizations as they seek shared understanding, adjust operations as necessary, and achieve objectives. This occurs in an environment where an enemy force is seeking to disrupt the friendly force's ability to effectively command and control its forces.

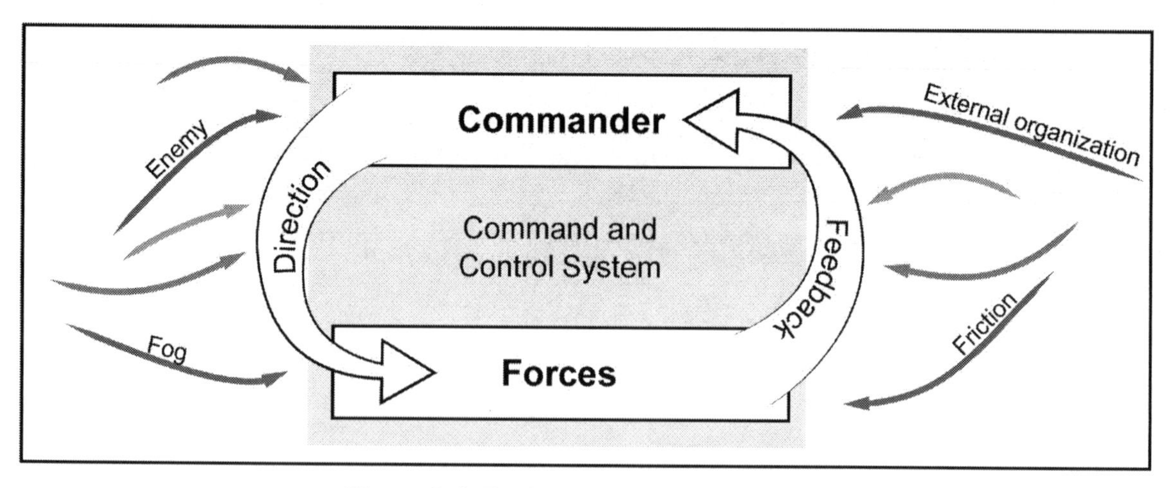

Figure 3-1. Reciprocal nature of control

DIRECTION

3-9. A key element of control is direction. To direct means to communicate information related to a decision that initiates and governs actions of subordinate and supporting units. Commanders, through their command and control system, direct subordinates by establishing objectives, assigning tasks, and providing instruction on how forces will cooperate to accomplish the mission.

3-10. The primary means for communicating direction include plans and orders. Mission orders (a technique for developing plans and orders) focus subordinates on what to do and why to do it without prescribing exactly how to do it. (See chapter 1 for more information on mission orders.) Other key tools for providing direction include execution matrices, the decision support template, and control measures. Control measures are directives to subordinate units that assign responsibilities, coordinate fires and maneuver, and control operations. Commanders establish control measures to aid cooperation among forces while minimizing restrictions on freedom of action as much as possible. (See paragraphs 3-40 to 3-45 for a detailed discussion of control measures.)

FEEDBACK

3-11. Feedback is information commanders receive during operations. Commanders use feedback to compare the actual situation with the plan and then decide if the plan requires any changes or adjustments. Feedback takes many forms, including information, knowledge, experience, and wisdom. Feedback comes from many sources: subordinates, higher headquarters, or adjacent, supporting, and supported forces. It arrives continuously: before, during, or after operations. Feedback helps commanders and subordinates gain shared understanding. For feedback to be effective, it should identify any differences between the desired end state and the current situation. New information that conflicts with the expectations established during planning requires commanders and staffs to validate those expectations or revise them to reflect reality. This contributes to an accurate understanding that allows commanders to exploit fleeting opportunities, respond to developing situations, modify plans, or reallocate resources.

3-12. Feedback should not flow only from lower to higher headquarters; it should also flow from higher to lower headquarters. Normally information from higher echelons to lower echelon headquarters consists of information to adjust the subordinates' resources, plans, or missions. Multidirectional information flow produces shared understanding between higher commanders and subordinate forces that supports exercise of mission command.

3-13. Effective commanders seek feedback from subordinates who are comfortable providing both positive and negative reports. Commanders whose command climates make subordinates reluctant to share bad news are likely to be poorly informed and operate from faulty assumptions that put operations at risk.

INFORMATION

> *Many intelligence reports in war are contradictory; even more are false; and most are uncertain....reports turn out to be lies, exaggerations, errors, and so on.*
>
> Carl von Clausewitz

3-14. Operations produce large amounts of information. While much of this information may be important to the staff or the conduct of operations, it may not be relevant information for the commander. ***Relevant information*** **is all information of importance to the commander and staff in the exercise of command and control.** Relevant information provides the basis for creating and maintaining the COP, and it is the basis for achieving situational understanding. Relevant information also facilitates a commander's decision making and the ability to provide timely orders and guidance.

3-15. Commanders assess information in the context of a particular situation. For example, in some situations, information that is somewhat incomplete or imprecise may be better than no information at all, especially when time for execution is limited. However, effective commanders take action with their staffs and subordinates to reduce the likelihood of receiving inaccurate, late, or unreliable information, which is of no value when making decisions. They do this by setting expectations, conducting training, and providing education.

3-16. Staffs provide commanders and subordinates information relevant to their operational environment and the progress of operations. They use operational variables (political, military, economic, social, information, infrastructure, physical environment, and time—known as PMESII-PT) and mission variables (mission, enemy, terrain and weather, troops and support available, time available and civil considerations—known as METT-TC) as major subject categories to group relevant information. (See FM 6-0 for discussion of the operational and mission variables.)

Operational Variables

3-17. Commanders and staffs analyze and describe an operational environment in terms of eight interrelated operational variables: political, military, economic, social, information, infrastructure, physical environment, and time (known as PMESII-PT). The operational variables are fundamental to developing a comprehensive understanding of an operational environment. Table 3-1 provides a brief description of each variable.

Table 3-1. Operational variables

Variable	*Description*
Political	Describes the distribution of responsibility and power at all levels of governance—formally constituted authorities, as well as informal or covert political powers
Military	Explores the military and paramilitary capabilities of all relevant actors (enemy, friendly, and neutral) in a given operational environment
Economic	Encompasses individual and group behaviors related to producing, distributing, and consuming resources
Social	Describes the cultural, religious, and ethnic makeup within an operational environment and the beliefs, values, customs, and behaviors of society members
Information	Describes the nature, scope, characteristics, and effects of individuals, organizations, and systems that collect, process, disseminate, or act on information
Infrastructure	Is composed of the basic facilities, services, and installations needed for the functioning of a community or society
Physical environment	Includes the geography and manmade structures, as well as the climate and weather in the area of operations
Time	Describes the timing and duration of activities, events, or conditions within an operational environment, as well as how the timing and duration are perceived by various actors in the operational environment

Mission Variables

3-18. Mission variables describe characteristics of the area of operations, focusing on how they might affect a mission. Incorporating the analysis of the operational variables into the mission variables ensures Army leaders consider the best available relevant information about conditions that pertain to the mission. Using the operational variables as a source of relevant information for the mission variables allows commanders to refine their situational understanding of their operational environment and to visualize, describe, direct, lead and assess operations. Table 3-2 provides a brief description of each of the mission variables. (See FM 6-0 for detailed discussion of the operational and mission variables.)

Table 3-2. Mission variables

Variable	*Description*
Mission	Commanders and staffs view all of the mission variables in terms of their impact on mission accomplishment. The mission is the task, together with the purpose, that clearly indicates the action to be taken and the reason therefore. It is always the first variable commanders consider during decision making. A mission statement contains the "who, what, when, where, and why" of the operation.
Enemy	The second variable to consider is the enemy's dispositions (including organization, strength, location, and tactical mobility), doctrine, equipment, capabilities, vulnerabilities, and probable courses of action.
Terrain and weather	Terrain and weather analysis are inseparable and directly influence each other's impact on military operations. Terrain includes natural features (such as rivers and mountains) and manmade features (such as cities, airfields, and bridges). Commanders analyze terrain using the five military aspects of terrain expressed in the memory aid OAKOC: observation and fields of fire, avenues of approach, key and decisive terrain, obstacles, cover and concealment. The military aspects of weather include visibility, wind, precipitation, cloud cover, temperature, and humidity.
Troops and support available	This variable includes the number, type, capabilities, and condition of available friendly troops and support. These include supplies, services, and support available from joint, host nation, and unified action partners. They also include support from civilians and contractors employed by military organizations, such as the Defense Logistics Agency and the Army Materiel Command.
Time available	Commanders assess the time available for planning, preparing, and executing tasks and operations. This includes the time required to assemble, deploy, and maneuver units in relationship to the enemy and conditions.
Civil considerations	***Civil considerations*** **are the influence of manmade infrastructure, civilian institutions, and attitudes and activities of the civilian leaders, populations, and organizations within an area of operations on the conduct of military operations.** Civil considerations comprise six characteristics, expressed in the memory aid ASCOPE: areas, structures, capabilities, organizations, people, and events.

3-19. Commanders determine information requirements and set information priorities. They avoid requesting excessive amounts of information, which may reduce the staffs' chances of finding what is actually important in a particular situation. The quest for information is time consuming; commanders who demand complete information place unreasonable burdens upon subordinates. Subordinates pressured to worry over every detail rarely have the desire to exercise initiative. At worst, excessive information demands corrupt the trust required for a mission command approach. Commanders describe the relevant information they need to inform decision making by establishing CCIRs.

Commander's Critical Information Requirement

It is in the minds of commanders that the issue of battle is really decided.

B.H. Liddell Hart

3-20. A *commander's critical information requirement* is an information requirement identified by the commander as being critical to facilitating timely decision making (JP 3-0). Commanders designate an information requirement as a CCIR based on likely decisions and their visualization of the operation. CCIRs help to bring clarity to large volumes of information. Always promulgated by a plan or order, commanders limit the number of CCIRs to focus the efforts of limited collection assets. The fewer the CCIRs, the easier it is for staffs to remember, recognize, and act on each one. CCIRs should change with the situation over time. Commanders add and delete them throughout an operation based on the information needed for specific decisions. Once approved, a CCIR falls into one of two categories: priority intelligence requirements and friendly force information requirements.

3-21. A *priority intelligence requirement* is an intelligence requirement that the commander and staff need to understand the threat and other aspects of the operational environment. (JP 2-01). Priority intelligence requirements identify the information about an enemy force and other aspects of the operational environment that a commander considers most important to the plan or decisions. Intelligence about civil considerations may be as critical as intelligence about enemy forces. Intelligence officers manage priority intelligence requirements for commanders as part of the intelligence process.

3-22. A *friendly force information requirement* is information the commander and staff need to understand the status of friendly force and supporting capabilities (JP 3-0). Friendly force information requirements identify the information about the mission, troops and support available, and time available for friendly forces that the commander considers most important to the plan or decisions. In coordination with staffs, the operations officers manage friendly force information requirements for commanders.

3-23. Commanders also describe information they want protected as essential elements of friendly information. An ***essential element of friendly information* is a critical aspect of a friendly operation that, if known by a threat would subsequently compromise, lead to failure, or limit success of the operation and therefore should be protected from enemy detection.** Although essential elements of friendly information are not CCIRs, they have the same priority. Essential elements of friendly information establish elements of information to protect rather than ones to collect. Their identification is the first step in the operations security process and central to the protection of information.

3-24. Commanders cannot recognize all their information requirements. There is information that results from an extraordinary event, an unseen opportunity, or a new threat. This is exceptional information—specific and immediately vital information that directly affects the success of the current operation. It may directly affect mission accomplishment or survival of the force, and usually reveals the need for a decision. It would have been a CCIR if it had been foreseen. Therefore, it is treated as a CCIR and is reported to the commander immediately. Identifying exceptional information requires initiative from subordinate commanders and staffs, shared understanding of the situation, and a thorough understanding of the commander's intent. It also requires professional judgment; if there is doubt it is better to report.

3-25. Commanders can neither make decisions nor act to implement them without information. The amount of information available makes managing information and turning it into effective decisions and actions critical to success during operations. Since effective command and control depends on getting relevant information to the right person at the right time, knowledge management, information management, and foreign disclosure are crucial to command and control.

Knowledge Management

He who wars walks in a mist through which the keenest eye cannot always discern the right path.

Sir William Napier

3-26. ***Knowledge management* is the process of enabling knowledge flow to enhance shared understanding, learning, and decision making**. Knowledge flow refers to the ease of movement of knowledge within and among organizations. Knowledge must flow to be useful. The purpose of knowledge management is to align people, processes, and tools within the organizational structure and culture to achieve shared understanding. This alignment improves collaboration and interaction between leaders and subordinates. Knowledge management leads to better decisions and increases flexibility, integration, and synchronization. Sound knowledge management practices include—

- Collaboration among personnel at different locations.
- Rapid knowledge transfer between units and individuals.

3-27. Knowledge management provides the methods and means to efficiently share knowledge and distribute relevant information where and when it is needed. Knowledge management organizes, applies, collects, codifies, and exchanges information between people. It seeks to align people and processes with appropriate tools to help units learn, adapt, and improve mission performance.

3-28. Knowledge management is supported by four tasks that bring an organization closer to situational and shared understanding. The four knowledge management tasks are creating knowledge, organizing knowledge, applying knowledge, and transferring knowledge. (See ATP 6-01.1 for more information on knowledge management.) Two important aspects of knowledge management are information management and foreign disclosure.

3-29. ***Information management* is the science of using procedures and information systems to collect, process, store, display, disseminate, and protect data, information, and knowledge products.** Information management supports, underpins, and enables knowledge management. The two are linked to facilitate understanding and decision making. Information management is a technical discipline that involves the planning, storing, manipulating, and controlling of information throughout its life cycle in support of the commander and staff. Information management employs both staff management and processes to make information available to the right person at the right time. Information management provides a structure so commanders and staffs can process and communicate relevant information and make decisions. Effective information management contributes to the knowledge management tasks of knowledge creation and supports shared understanding for all unit members.

3-30. Generally, information management relates to the tasks of collection, processing, display, storage, distribution, and protection of data and information. In contrast, knowledge management uses information to create, organize, apply, and transfer knowledge to support achieving understanding, making decisions, and ultimately taking effective action.

3-31. Information management provides the timely and protected distribution of relevant information to commanders and staff elements. It supports and is a component of knowledge management. (See FM 6-0.2 for a discussion of information management.)

3-32. Foreign disclosure is a critical part of interoperability with multinational partners. Conducting operations with unified action partners affects the way the force collects and disseminates information. The disclosure of classified and controlled unclassified information to foreign representatives is governed by policy and regulations. Keeping as much information unclassified as possible improves interoperability, operational effectiveness, and trust.

3-33. Determining what information or intelligence may be disclosed is based on the policies, directives, and laws that govern national disclosure policy and the release of classified information. While conducting operations, commanders and staffs ensure they know other nations' positions on intelligence sharing and ensure that intelligence is shared to the degree possible, especially if required for mission accomplishment and force protection. Early information sharing during planning ensures that unified action partner requirements are clearly stated, guidance supports the commander's intent, and the force uses procedures supportable by other nations. (See AR 380-10 for more discussions on foreign disclosure.)

Communication

3-34. Commanders and staffs disseminate and share information among people, elements, and places. Communication is more than the simple transmission of information. It is a means to exercise control over forces. Communication links information to decisions and decisions to action. Communication among the parts of a command supports their coordinated action. Effective commanders do not take communication for granted. They use multidirectional communication and suitable communication media to achieve objectives. Commanders choose appropriate times, places, and means to communicate. They use face-to-face talks, written and verbal orders, estimates and plans, published memos, electronic mail, and other methods of communication appropriate for a particular situation.

3-35. Communication has an importance far beyond exchanging information. Commanders and staffs continuously communicate to learn, exchange ideas, and create sustained shared understanding. Information needs to flow up and down the chain of command as well as laterally to adjacent units and organizations.

Separate from the quality or meaning of information exchanged, communication strengthens bonds within a command. It is an important factor in building trust, cooperation, cohesion, and mutual understanding.

3-36. Humans communicate verbally by what they say and in their manner of speaking. They also communicate nonverbally with gestures and body language. Commanders pay attention to verbal and nonverbal feedback to ascertain the effectiveness of their communication and the manner in which it is received. Commanders should conduct face-to-face talks with their subordinates to ensure subordinates fully understand them. This does not mean they do not keep records of information communicated or follow-up with written documentation. Records are important as a means of affirming understanding and for later study and critique. Records support understanding over time, whereas memory may distort or even omit elements of the information required or passed.

3-37. In many cases, commanders are tempted to rely too much on written communications; especially email. Email messages, written papers, briefs, and directives do not have the same impact as oral orders, consultations, and briefings. Staffs possess the ability to produce vast amounts of documents; however, just because the capability exists does not mean it should be used. Quality communication is superior to quantity for enabling a mission command approach to command and control.

3-38. Commanders should assume that communications will be disrupted during operations. Commanders' intent and orders should be written in a way that enables achieving objectives when communication is intermittent and situational awareness is problematic. Mission orders and application of the mission command approach to command and control mitigates the need for continuous communication.

Channels

3-39. Information normally moves throughout a force along various transmission paths or channels. Commanders and staffs transfer information horizontally and vertically. Structure, in the form of command and support relationships, establishes channels that streamline information dissemination by ensuring the right information passes promptly to the right people. Commanders and staffs communicate through three channels—command, staff, and technical:

- Command channels are direct chain-of-command transmission paths. Commanders and authorized staff officers use command channels for command-related activities.
- Staff channels are staff-to-staff transmission paths between headquarters and are used for control-related activities. Staff channels transmit planning information, status reports, controlling instructions, and other information to support mission command. The intelligence and sustainment networks are examples of staff channels.
- Technical channels are the transmission paths between two technically similar units, offices, or staff sections that perform a technical function requiring special expertise or control the performance of technical functions. Technical channels are typically used to control performance of technical functions. They are not used for conducting operations or supporting another unit's mission. An example is network operations. The activities for the operation, management, and control of communications transport is routinely performed by network operations control centers.

3-40. Crosstalk between subordinate commanders can transfer information and lead to decision making without the higher echelon commander becoming involved, except to affirm, either positively or through silence, the decisions or agreements of subordinates. However, commanders must train their subordinates to crosstalk, so they can quickly and competently exchange information, create shared understanding, and make and implement decisions.

> **Crosstalk in the Desert-VII Corps in the Gulf War**
>
> On the morning of 17 January 1991, the day after the start of U.S. Central Command's major air operations against Iraq, the VII Corps Commander, LTG Frederick M. Franks Jr., was with the 1st Infantry Division as it honed tank and Bradley gunnery skills in the desert of Saudi Arabia. While there, he received a spot report from BG John Landry, corps chief of staff, over FM radio: "55 Iraqi tanks have crossed the Kuwaiti Border, heading southwest toward Hafir al-Batin and are engaging Egyptian coalition forces in what may be the beginnings of an Iraqi preemptive strike."
>
> Within seconds, COL Johnnie Hitt, commander of the corps' 11th Aviation Brigade, entered the net indicating he had monitored the report and alerted two Apache battalions that could respond in 30 minutes if necessary. At the same time, COL Don Holder, commander of the 2nd Armored Cavalry Regiment, the corps unit closest to the reported enemy force, called to notify LTG Franks that he had issued orders for 1st Squadron to send a unit forward to recon and make contact with enemy forces. Those were the immediate and correct actions taken by commanders as a result of eavesdropping on the command net and having the confidence to act—confidence developed through training, teamwork, and trust among the key players of the VII Corps team.

Structure

3-41. Organizational structure helps commanders communicate information and exercise control. Structure refers to a defined organization that establishes relationships, information flow, and guides interactions among elements. It also includes procedures for coordinating among an organization's groups and activities. Commanders establish control with a defined organization. Structure is both internal (such as a command post) and external (such as command and support relationships among subordinate forces). (See ATP 6-0.5 for information on organizing Army command post operations and FM 3-0 for information on command and support relationships.) The most basic organization in control is a hierarchy. In military terms, this relationship is between the commander and staff, and subordinate forces.

CONTROL MEASURES

3-42. Commanders use control measures to assign responsibilities, coordinate fire and maneuver, and control operations. **A *control measure* is a means of regulating forces or warfighting functions.** Control measures provide control without requiring detailed explanations. Control measures help commanders direct actions by establishing responsibilities and limits that prevent subordinate unit actions from impeding one another. They foster coordination and cooperation between forces without unnecessarily restricting freedom of action.

3-43. Control measures may be detailed (such as an operation order) or simple (such as a checkpoint). Control measures include, but are not limited to—

- Plans and orders.
- Laws and regulations.
- Unit standard operating procedures.

3-44. Some control measures are graphic. A ***graphic control measure* is a symbol used on maps and displays to regulate forces and warfighting functions.** Graphic control measures are always prescriptive. They include symbols for boundaries, fire support coordination measures, some airspace control measures, air defense areas, and minefields. Commanders establish them to regulate maneuver, movement, airspace use, fires, and other aspects of operations. In general, all graphic control measures should relate to easily identifiable natural or man-made terrain features. (See ADP 1-02 for illustrations of graphic control measures and rules for their use.)

3-45. Control measures are established under a commander's authority; however, commanders may authorize staff officers and subordinate leaders to establish them. Commanders may use control measures for several purposes: to assign responsibilities, require synchronization between forces, impose restrictions, or establish guidelines to regulate freedom of action. Certain control measures belong to the commander alone, and may not be delegated. These include the commanders' intent, unit mission statement, planning guidance, and CCIRs and essential elements of friendly information.

3-46. Good control measures foster freedom of action, decision making, initiative, and reporting during operations. Commanders tailor their use of control measures to conform to the higher echelon commander's intent. They also consider the mission, terrain, and amount of authority delegated to subordinates. Effectively employing control measures requires commanders and staffs to understand their purposes and ramifications, including the permissions or limitations imposed on subordinates' freedom of action and initiative. Each measure should have a specific purpose: mass the effects of combat power, synchronize subordinate forces' operations, minimize the possibility of fratricide, or comply with the law of armed conflict.

3-47. The most important control measure is the boundary. Boundaries define the area of operations assigned to a commander. Commanders have full freedom of action to conduct operations within the boundaries of their area of operations unless the order establishing the area of operations includes constraints.

Control in Austerlitz

Napoleon's La Grande Armée of 1805 had spent two years training along the coast of the English Channel to invade England. On 3 September 1805, after the Third Coalition formed, Napoleon moved that army against the first opposing force that presented itself. His desired end state was to defeat it before the rest of the coalition forces could join the campaign. Napoleon marched east with 200,000 men. He defeated an Austrian army at Ulm in Bavaria by 20 October 1805 and pursued an approaching Russian army down the Danube River toward Vienna.
On 23 November, he halted his pursuit east of Brunn (present-day Brno, Czech Republic) near the village of Austerlitz, 700 miles from the Channel coast.

The Russian army had joined another Austrian army to form a force that numbered 85,000 to Napoleon's 53,000. Napoleon decided to entice the coalition force to attack him before others could reinforce it. He displayed his weakness in numbers, which he let the coalition commanders see, and withdrew his main body from the Pratzen Heights, key terrain in the area he had selected for battle. The coalition force occupied that terrain on 30 November and prepared for battle. Napoleon had two corps moving to reinforce his main body, increasing its strength to 73,000 before the battle: one joined him on 1 December; the other, with 50 hours to march 80 miles from Vienna, would not arrive until the day of the battle. (See figure 3-2 on page 3-12.)

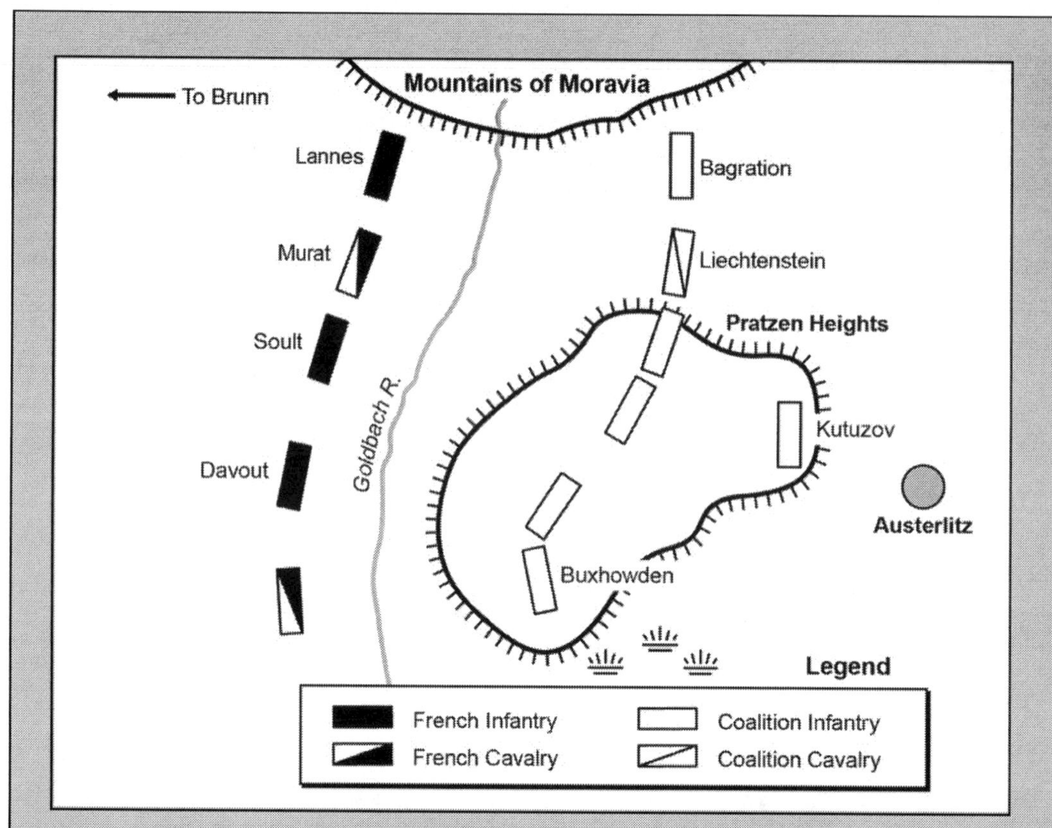

Figure 3-2. Map of Austerlitz, the initial situation

Napoleon planned to show weakness on his right flank, which was held by a single division. This display would encourage the coalition commanders to attack there. He would hold on his left flank and attack the coalition center, where the coalition had taken forces to carry out its attack on his right. With his forces attacking in the center, Napoleon could either roll up the coalition forces attacking to his left, or more decisively envelop those attacking his right. Key to this was the timely (for him), unexpected (for the coalition) arrival of the corps from Vienna (under Marshal Davout). It reinforced his right as the coalition attack began.

The attack against Napoleon's right began at 0600 on 2 December and had intensified by 0700. A coalition attack against his left also threatened but had not yet commenced. Davout's lead forces reinforced the French right by 0700, and the fight there continued for the next two hours; a French force of 10,600 occupied a coalition force of nearly 50,000. By 0800, Napoleon, from his command post, could directly observe most of the coalition force moving against his right. The Pratzen Heights, key terrain that he had given up to entice the coalition commanders to give battle, was now uncovered. By 0830, Napoleon had also received reports about the tenacious, successful fight of his right and that his left was still secure.

Hidden from coalition view but within striking distance of the key terrain were two French divisions, 16,000 men and 16 guns, under command of "the finest maneuverer in Europe," Marshal Soult. Through the initial fight, Soult chafed to commence his attack, but Napoleon restrained him. At 0845, Napoleon turned to Soult and asked, "How long will it take you to move your divisions to the top of Pratzen Heights?" "Less than 20 minutes, sire," Soult answered. "Very well, we'll wait another quarter of an hour," decided Napoleon. By then, Napoleon knew that a coalition force had begun attacking his left. At 0900, Napoleon turned to Soult and directed him to attack: "One sharp blow and the war's over." By 0930, Soult had taken the Pratzen Heights and was well on the way to securing it.

The French left now also attacked the coalition right with coordinated infantry and cavalry actions under Marshals Lannes and Murat. By noon, this French shaping operation drove the coalition right back four miles, making it unable to move against the decisive operation on the Pratzen. Stationing himself in the center, Napoleon remained informed of events on both flanks but did not direct subordinate actions. Napoleon's situational understanding and ability to regulate his forces were enhanced by a semaphore (signal flag) station at his command post and relay stations throughout the area of operations. (See figure 3-3.)

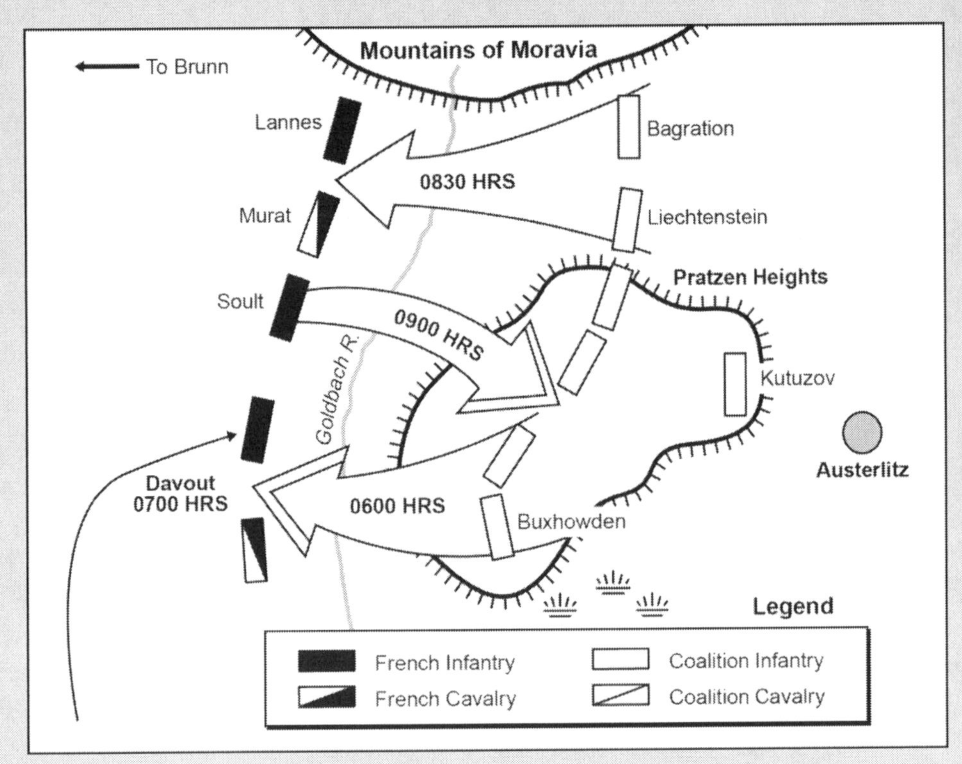

Figure 3-3. Map of Austerlitz operations

Soult's assault of the Pratzen only began the struggle in the center. The Russian commander, Marshal Kutuzov, recognized the danger and recalled forces from attacking the French right to counter Soult's attack. The battle against this counterattack began about 1000 and continued through 1100. By noon, Napoleon had moved his command post and his reserve up to the Pratzen. The Russian Imperial Guard mounted another counterattack against the center at 1300, but the well-positioned French reserves, in coordination with Soult's forces, defeated them by 1400 after much hard fighting.

This left Napoleon with the initiative to envelop either coalition wing. Napoleon had an accurate' situational understanding. He knew the coalition right could neither intervene against him nor support the coalition center. He was also aware that nearly half of the coalition force still engaged the French right, with a lake to their south. Accordingly, he directed his center to wheel south (to its right), taking the coalition left in the rear and destroying it. He left one corps in the center to secure the Pratzen Heights and his rear, while Soult's corps and Napoleon's Imperial Guard executed the envelopment to the south. By 1430, the coalition commander in the south recognized the peril to his force and directed its retreat. About half escaped the encirclement by 1500. Some of the encircled coalition forces attempted to escape over the frozen lake to the south, but French artillery fired at the ice, breaking it and cutting off that avenue, while drowning over 200 men. By 1500, the coalition right wing began to retreat as well, and by 1630, as dark fell, all firing stopped. The coalition army was destroyed, with over one-third of its force lost.

Napoleon's ability to employ the necessary level of control was major factor in this victory. It allowed him to move his Army across Europe and regulate his forces' execution of a complex scheme of maneuver while decisively engaged with an adaptive opponent. The timeliness of Napoleon's decisions, which were the result of significant preparations and thought, rendered his enemies' reactions progressively more irrelevant as the battle went on.

GUIDES TO EFFECTIVE CONTROL

...avoid taking "firm control" or a "tight rein" over the battle...these measures are likely to hold back the offensive during a penetration or pursuit and thus damage their chances of success.

Marshal of the Soviet Union, Mikhail N. Tukhachevskiy

3-48. The guides to effective control govern how commanders use the elements of control to accomplish missions. Effective control enables a command to adapt to change. Because of feedback, control is cyclic and continuous, not a series of discrete actions. It is a process of dynamic, interactive cooperation. Control begins in planning and continues throughout the operations process. The guides to effective control are—

- Allow subordinates maximum freedom of decision and action.
- Create, maintain, and disseminate the COP.
- Use common doctrinal procedures, graphics, and terms.
- Encourage flexibility and adaptability.

ALLOW SUBORDINATES MAXIMUM FREEDOM OF DECISION AND ACTION

3-49. Effective commanders impose minimum constraints on subordinates to enable freedom of action while meeting the overall intent. They exercise the control necessary to effect coordination and synchronization among subordinate and supporting forces. Commanders monitor this coordination and allocate available resources or shift priorities to support the actions of subordinate commanders. Allowing subordinates maximum freedom of action requires a mission command approach.

3-50. In most instances, lower echelon commanders have the clearest understanding of their own situations. They are generally better suited than higher echelon commanders to develop those situations. Even two or more subordinate commanders working together may solve a problem better and faster than the higher echelon commander. This type of coordination, involving direct communication among subordinate commanders, is critical for effective decentralized execution. Commanders emphasize lateral coordination at every opportunity.

3-51. Commanders should avoid establishing excessive limits on their subordinates' freedom of action. These limits may come in the form of overly detailed orders that inhibit initiative and force subordinates to refer large numbers of decisions to their higher headquarters. Excessive detail may be a result of emphasizing process or procedure rather than an outcome—directing things already addressed in doctrine or standard operating procedures.

3-52. Commanders limit information requests to that which is critical to decision making. Excessive requests for information may burden subordinates with reporting and distract them from executing their operation. They can also affect the requesting unit, because it must process the responses. One cause of excessive requests is the search for perfect situational understanding. Another stems from poor information management. No one can predict all information requirements before operations begin; however, commanders and staffs must balance new information requirements against the impact that finding and providing that information will have on subordinates' operations. Excessive or redundant reporting can create unnecessary stress or fatigue for subordinate units. This situation may result in subordinates failing to respond to an important request and depriving the higher commander of critical information needed to make decisions.

3-53. Commanders consider these items when deciding how to exercise control:

- Limit control measures to those necessary to effect essential coordination.
- Limit information requirements to the minimum needed to exercise command and control.
- Give subordinates as much leeway for initiative as possible.

Create, Maintain and Disseminate the Common Operational Picture

3-54. The COP is key to achieving and maintaining situational understanding. **The *common operational picture* is a display of relevant information within a commander's area of interest tailored to the user's requirements and based on common data and information shared by more than one command.** Although the COP is ideally a single display, it may include more than one display and information in other forms, such as graphical representations or written reports.

3-55. The COP facilitates collaborative planning and helps commanders at all echelons achieve shared situational understanding. Shared situational understanding allows commanders to visualize the effects of their decisions on other elements of the force and the overall operation. Mission command allows subordinates to use the COP in conjunction with the commander's intent to guide their exercise of disciplined initiative.

3-56. Commanders achieve situational understanding by applying judgment to the COP. Relevant information provides the basis for constructing the COP, and primarily consists of information which the staff provides through analysis and evaluation. Data and information from all echelons and shared among users create the COP. Some sources of this information include reports, running estimates, and information provided by liaison officers.

3-57. Maintaining an accurate COP is difficult for many reasons: delayed or inaccurate friendly and intelligence reporting, terrain data availability, and a constantly changing operational environment that change circumstances, often in unforeseen ways. Fog and friction often degrade the accuracy of the COP, or render all or part of it incorrect do to latency, enemy deception plans, or incorrect reporting. Staffs, based on guidance from the commander, must work to rapidly and accurately portray the meaning and the necessary level of information which help the commander maintain situational understanding and update his visualization. Staffs should only display information that is relevant to the commander's decision-making, and avoid overloading their commander with unnecessary details.

3-58. Units continually refine the COP during operations based on information they receive. By collaborating and sharing relevant information and tailoring it to their needs, separate echelons create their own COPs that show what their commanders need to know, as the situation requires.

3-59. Command posts draw on a common set of relevant information within a shared database to create a digital COP. When a digital COP is not possible due to the operational environment or network interruptions, command posts employ an analog COP. Creation of a COP is done manually with physical maps and overlays that require training to employ effectively. The use of overlays or gathering subordinates around a common map or graphic are examples of applying the COP concept in an analog command post.

3-60. Maintaining a COP between units within any organization is a challenge. Maintaining a COP between different countries in a multinational environment is more of a challenge. The difficulty level varies based on language differences and technical compatibility of systems. Often unified action partners will not have the technical capability or compatible systems to create and share a digital COP. Commander's must recognize and plan for this possibility.

Use Doctrinally Correct Terms and Graphics

A doctrine of war consists first in a common way of objectively approaching the subject; second, in a common way of handling it.

Ferdinand Foch; *Precepts*, 1919

3-61. Language used when communicating should be simple, clear, and easily understood. An understanding of common doctrinal procedures, graphics, and terms contributes to the simplicity and clarity essential to mutual understanding and shortens the amount of explicit communication needed to convey or explain an order or plan. Doctrinal terms and graphics enable shared understanding by communicating in a commonly understood way. However, there are situations where staffs may need to create nonstandard graphics or modify existing graphics to portray the environment, an adaptive opponent, or other elements. They should do this only when standard graphics are unsuitable, and they must ensure that subordinates and adjacent units understand nonstandard terms and graphics.

3-62. Doctrine clearly distinguishes between descriptive and prescriptive information. Most doctrine is descriptive; it must be applied with judgment in the context of a particular situation. Unthinking adherence to every aspect of doctrine in inappropriate situations is not congruent with a mission command approach. There are some reasons that Army forces must apply some aspects of doctrine prescriptively—done without deviation. Prescriptive doctrine derives from the need to—

- Adhere to the Army Ethic, laws of war, national law, the Uniform Code of Military Justice, and often Army regulations.
- Precisely use terms, symbols, and the language of the profession to ensure a common understanding.
- Adhere to control measures to ensure coordination, ensure synchronization, and prevent fratricide.
- Use report, message, and order formats to ensure information is reported rapidly, accurately, and in a commonly understood manner.

Encourage Flexibility and Adaptability

3-63. Control allows organizations to respond to change, whether due to threat or friendly actions, or environmental conditions. The mission command approach provides flexibility and adaptability, allowing subordinates to recognize and respond effectively to emerging conditions and to correct for the effects of fog and friction. Control informed by a mission command approach provides information that allows commanders to base their decisions and actions on the results of friendly and opponent actions, rather than rigid adherence to the plan. Commanders seek to build flexibility and adaptability into their plans.

3-64. Control supports flexibility and adaptability in two ways. First, it identifies the need to change the plan. It does this through anticipating or forecasting possible opponent actions and by identifying unexpected variances—opportunities or threats—from the plan. This occurs throughout the operations process. Second, control helps commanders develop and implement options to respond to these changes in a timely manner. Flexibility and adaptability provided by the appropriate level of control reduces an opposing force's available options while maintaining or expanding friendly options. Effective control provides for timely action before opposing forces can accomplish their objectives, allowing for the modification of plans as the situation changes.

3-65. Instead of rigidly adhering to the plan, control focuses on information about emerging conditions. The mission command approach to control provides flexibility by—

- Allowing friendly forces to rapidly change their tasks, their task organization, or their plan in response to changing circumstances.
- Producing information about options to respond to changing conditions.
- Communicating the commander's decisions quickly and accurately.
- Providing for rapid reframing when the plan changes during execution.
- Allowing collaborative planning and execution to respond to the progress of operations.

CONCLUSION

3-66. Control is essential to the ability of commanders to counter the effects of enemy action, fog, and friction during operations, and it allows commanders to turn decisions into effective action. Knowledge management supports control by providing structure to communications and transforming information into knowledge in support of situational understanding and decision making. Commanders exercise control through the operations process and the command and control system discussed in chapter 4.

This page intentionally left blank.

Chapter 4
The Command and Control System

Staff systems and mechanical communications are valuable, but above and beyond them must be the commander; not as a disembodied brain linked to his men by lines of wire and waves of ether; but as a living presence, an all pervading visible personality...

General George S. Patton, Jr

This chapter expands on the command and control system that performs the functions necessary to exercise command and control. First, it defines the command and control system and its purpose. It then describes the individual components of the command and control system, followed by a discussion of organizing for command and control. Finally, this chapter concludes with a discussion of design considerations when establishing a command and control system.

COMMAND AND CONTROL SYSTEM DEFINED

4-1. Commanders cannot exercise command and control alone. Even at the lowest levels, commanders need support to exercise command and control. At every echelon of command, each commander has a command and control system to provide that support. The command and control system is the arrangement of people, processes, networks, and command posts that enable commanders to conduct operations.

4-2. The command and control system consists of all the resources used to support command and control and enhances the commander's ability to conduct operations. Commanders organize a command and control system to—

- Support the commander's decision making.
- Collect, create, and maintain relevant information and prepare products to support the commander's and leaders' understanding and visualization.
- Prepare and communicate directives.

4-3. To provide these three overlapping functions, commanders must effectively locate, design, and organize the four components of their command and control system (depicted in figure 4-1): people, processes, networks, and command posts.

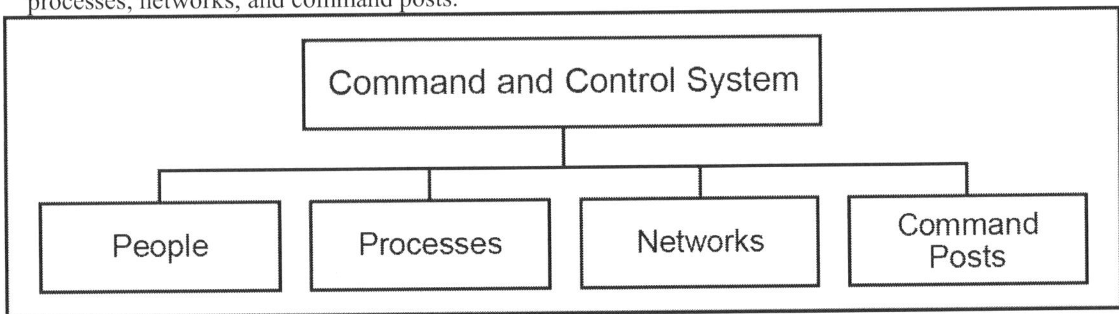

Figure 4-1. Components of a command and control system

PEOPLE

4-4. The most important component of the command and control system is people—those who assist commanders and exercise control on their behalf. An effective command and control system accounts for the characteristics and limits of human nature. Simultaneously, it exploits and enhances uniquely human skills. People dedicated to the command and control system include commanders, seconds in command, command sergeants major, staffs, and liaison officers.

Commanders

4-5. Where the commander locates within the area of operations, and at what time, are important considerations for effective command and control. No standard pattern or simple prescription exists for command presence; different commanders lead differently. Commanders balance their time among the command post and staff, subordinate commanders, forces, and other organizations to make the greatest contribution to success. (See chapter 2 for more discussion on the location of the commander.)

4-6. Command posts serve as the focus for information exchange, planning, and analysis. They provide commanders direct access to the staff and allow them to communicate with superior, subordinate, and supporting commanders. While at a command post, commanders provide vital face-to-face guidance to staff members when developing plans and controlling operations. By moving to the locations of subordinates or to critical points in an area of operations, commanders can better assess and influence the state of operations. They can personally gauge the condition of their units and leaders and consult directly with subordinate commanders performing critical tasks. By being forward, commanders can also motivate subordinates through personal example.

Seconds in Command

4-7. At all levels, the second in command is the commander's principal assistant. The second in command may be a deputy commander, an assistant commander, or the executive officer. Commanders determine the duties and responsibilities of their deputy and assistant commanders. These duties and responsibilities are formally declared and outlined in a memorandum or standard operating procedure signed by the commander. Usually, at company through brigade echelons, the executive officer is the second in command. In a theater army, corps, or division, the deputy or assistant commander extends the commander's span of control in areas and functions that the commander designates

4-8. Deputy or assistant commanders normally do not have staffs. When they have specific responsibilities, the headquarters staff assists them as their commander prescribes. Deputy or assistant commanders may give orders to the staffs within the authority the commander delegates to them. They may go to the chief of staff at any time for staff assistance. If a deputy or assistant commander needs to form a team for a specific purpose, the commander may form one from headquarters elements or subordinate units, or make a subordinate unit's headquarters available.

4-9. In organizations with more than one deputy or assistant, the commander will designate which one is the second in command. Delegating authority to the seconds in command reduces the burden on commanders and allows them to focus on particular areas or concerns while their seconds in command concentrate on others. Normally, commanders delegate authority to seconds in command to act in their name for specific functions and responsibilities.

4-10. A second in command has important responsibilities in these circumstances:

- Temporary absence of the commander.
- Succession of command.
- Delegation of authority.
- Deputies of joint and multinational forces.

Temporary Absence of the Commander

4-11. Officers who are second in the chain of command may assume duties as delegated, either explicitly or by standard operating procedures, when the commander is temporarily absent from the command post or resting. Lack of sleep can impair judgment and creative thinking capabilities. A commander's sleep plan should include delegating authority to the second in command during selected times to give the commander time to sleep. Commanders may also be absent from the command temporarily. In this case, the second in command may assume command temporarily and make decisions that continue operations in accordance with the commander's intent and policies.

Succession of Command

When placed in command-take charge...

GEN H. Norman Schwarzkopf

4-12. Commanders may be killed, wounded, medically incapacitated, or relieved of command. In these situations, the second in command normally assumes command. At brigade and lower echelons, executive officers normally assume command. At higher echelons, deputy or assistant commanders may not be senior to subordinate unit commanders. In this case, the operations order specifies succession of command, and the second in command exercises command until the designated successor assumes command. However, commanders may designate a second in command who is junior to subordinate commanders as their successor in command. (See AR 600-20 for regulatory guidance.)

4-13. Seconds in command must be able to assume command at any time. They must stay informed of the situation. Commanders inform their second in command of any changes to their visualization or intent. The chief of staff keeps the second in command informed of staff actions. Commanders continually train their seconds in command for command at their level.

Delegation of Authority

4-14. Delegating authority to the second in command reduces the burden of commanders' responsibilities and allows them to focus on particular areas or concerns while their seconds in command concentrate on others. Normally, commanders delegate authority to their seconds in command to act in their name for specific fields of interest and responsibility. Doing this decentralizes decision making while allowing the commander to keep overall command.

Deputies of Joint and Multinational Forces

4-15. When an Army headquarters serves as the headquarters of a joint or multinational force, appointing a deputy commander from another service or a multinational partner is often appropriate. These deputy commanders may also exercise command over forces of their Service or nation. They can serve as important advisers to the Army commander. They can facilitate understanding among participating Service or national forces. In this case, succession of command depends on joint and multinational doctrine, law, or international agreement.

Command Sergeants Major

4-16. The command sergeant major is the senior noncommissioned officer of the command at battalion and higher levels. Command sergeants major carry out policies and enforce standards for the performance, individual training, and conduct of enlisted Soldiers. They give advice and initiate recommendations to commanders and staffs in matters pertaining to enlisted Soldiers. In operations, commanders employ their command sergeant majors throughout the area of operations to extend command influence, assess morale of the force, and assist during critical events. Company first sergeants and platoon sergeants perform similar functions at company and platoon levels.

Staffs

A lazy commander, if he is brilliant, may succeed; but a lazy staff officer is a menace.

Lt.-Col. Simonds, Commandant, Canadian Junior War Staff Course

4-17. Staffs support commanders in making and implementing decisions and in integrating and synchronizing combat power. Effective staffs multiply a unit's effectiveness. They provide timely and relevant information and analysis, make estimates and recommendations, prepare plans and orders, assist in controlling operations, and assess the progress of operations for the commander. Primary responsibilities of any staffs are to—

- Support the commander.
- Assist subordinate commanders, staffs, and units.
- Inform units and organizations outside the headquarters.

Support the Commander

4-18. Staffs support commanders in understanding, visualizing, and describing the operational environment; making and articulating decisions; and directing, leading, and assessing military operations. Staffs make recommendations and prepare plans and orders for their commander. Staff products consist of timely and relevant information and analysis, such as that found in their running estimates. Staffs use knowledge management to extract that information from the vast amount of available information. Staffs synthesize this information and provide it to commanders in the form of running estimates to help commanders build and maintain their situational understanding.

4-19. Staffs also prepare and disseminate information to subordinates for execution to assist commanders in controlling operations. While commanders often personally disseminate their commander's intent and planning guidance, they rely on their staffs to communicate the majority of their guidance in the form of plans and orders. Staffs must communicate their commander's decisions, and the intentions behind them, efficiently and effectively throughout the force.

4-20. Staffs support and advise their commander within their area of expertise. While commanders make key decisions, they are not the only decision makers. Trained and trusted staff members, given decision making authority based on the commander's intent, free commanders from routine decisions. This enables commanders to focus on key aspects of operations.

Assist Subordinate Commanders, Staffs, and Units

4-21. Effective staffs establish and maintain a high degree of coordination and cooperation with staffs of higher echelon, lower echelon, supporting, supported, and adjacent units. Staffs help subordinate headquarters understand the larger context of operations. They do this by first understanding their higher echelon headquarters' operations and commander's intent, and nesting their own operations with their higher headquarters. They then actively collaborate with subordinate commanders and staffs to facilitate a shared understanding of the operational environment. Examples of staffs assisting subordinate units include performing staff coordination, staff assistance visits, and staff inspections.

Inform Units and Organizations Outside the Headquarters

4-22. Staffs keep their units well informed. Staffs also keep adjacent, coalition, allied, and civilian organizations informed with relevant information according to security classification and need to know. Staffs have an obligation to establish working relationships with unit members. As soon as a staff receives information and determines its relevancy, that staff passes that information to the appropriate headquarters. The key is relevance, not volume. Masses of data inhibit mission command by distracting staffs from relevant information. Effective knowledge management helps staffs identify the information commanders and staff elements need, and its relative importance.

4-23. Information should reach recipients based on their need for it. Sending incomplete information sooner is better than sending complete information too late to matter. When forwarding information, the sending staff highlights key information for each recipient and clarifies the commander's intent. Such highlighting and clarification assists receivers in analyzing the content of the information received in order to determine that information that may be of particular importance to the higher and subordinate commanders. The sending staff may pass information directly, include its analysis, or add context to it. Common, distributed databases can accelerate this function; however, they cannot replace the personal contact that adds perspective.

Common Staff Duties and Responsibilities

4-24. Staff members have specific duties and responsibilities associated with their area of expertise. However, all staff sections share a common set of duties and responsibilities:

- Advising and informing their commander.
- Building and maintaining running estimates.
- Providing recommendations.
- Preparing plans, orders, and other staff writing.
- Assessing operations.
- Managing information within their area of expertise.
- Identifying and analyzing problems.
- Conducting staff assistance visits.
- Performing risk management.
- Performing intelligence preparation of the battlefield.
- Conducting staff inspections.
- Conducting staff research.
- Performing staff administrative procedures.
- Exercising staff supervision over their area of expertise.

Characteristics of Good Staff Members

4-25. Good staff members understand how to effectively communicate with their commander, and they can discern what information is vital to their commander's ability to command and control. They seek a shared understanding of the operational environment with their commander and with the commanders of both higher and subordinate headquarters. This shared understanding includes the commander's visualization of the operational approach, including the commander's intent. Good staff members—

- **Are competent**. They are experts in doctrine and the processes and procedures associated with their branch or functional area, as well as the operations process. They understand the duties of other staff members enough to accomplish coordination both vertically and horizontally.
- **Bring clarity**. They are able to clearly articulate and effectively present information, orally, in writing, and visually. They help simplify problems in complex operational environments by explaining the meaning of information and not simply providing raw data to the commander.
- **Exercise candor**. They tell the commander what they believe, not what the commander wants to hear. They are willing to tell the commander both good and bad news. Any staff work eventually affects Soldiers, who execute staff officer recommendations approved by the commander. Staff officers never forget their recommendations affect Soldiers.
- **Exercise initiative**. They anticipate requirements rather than waiting for instructions. They anticipate what the commander needs to accomplish the mission and prepare answers to potential questions before they are asked.
- **Apply critical and creative thinking**. As critical thinkers, staff members discern truth in situations where direct observation is insufficient, impossible, or impractical. They determine whether adequate justification exists to accept conclusions as true, based on a given inference or argument. As creative thinkers, staff members look at different options to solve problems. They use proven approaches (drawing from previous similar circumstances) or innovative approaches (coming up with completely new ideas). In both instances, staff members use creative thinking to apply imagination and depart from the old way of doing things.
- **Are adaptive**. They recognize and adjust to changing conditions in the operational environment with appropriate, flexible, and timely actions. They rapidly adjust and continuously assess plans, tactics, techniques, and procedures.
- **Are flexible**. They avoid becoming overwhelmed or frustrated by changing requirements and priorities. Commanders may change their minds or redirect their commands after receiving additional information or a new mission and may not inform their staffs of the reason for a change. Staff members remain flexible and adjust to any changes. They set priorities when there are more tasks to accomplish than time allows. They learn to manage multiple commitments simultaneously.

- **Possess discipline and self-confidence**. They understand that all staff work serves the commander, even if the commander rejects the resulting recommendation. Staff members do not give a "half effort" even if they think the commander will disagree with their recommendations. Alternative and possibly unpopular ideas or points of view assist commanders in making the best possible decisions.
- **Are team players**. They cooperate with other staff members within and outside their headquarters. This practice contributes to effective collaboration and coordination.

Staff Relationships

4-26. Staff effectiveness depends in part on relationships of the staff with commanders and other staffs. Collaboration aids in developing shared understanding and visualization among staffs at different echelons. A staff acts on behalf of, and derives its authority from, its commander. Although commanders are the principal decision makers, individual staff officers make decisions within their authority based on broad guidance and unit standard operating procedures. Commanders insist on frank collaboration between themselves and their staff officers. A staff gives honest, independent thoughts and recommendations, so commanders can make the best possible decisions. Once their commander makes a decision, staff officers support and implement the commander's decision even if the decision differs from their recommendations.

4-27. Teamwork within a staff and between staffs produces the staff integration essential to synchronized operations. A staff works efficiently with complete cooperation from all staff sections. A force operates effectively in cooperation with all headquarters. Commanders and staffs foster this positive climate during training and sustain it during operations. However, frequent personnel changes and augmentation to their headquarters adds challenges to building and maintaining the team. While all staff sections have clearly defined functional responsibilities, none can operate effectively in isolation. Therefore, coordination is extremely important. Commanders ensure staff sections are properly equipped and manned. This allows staffs to efficiently work within their headquarters and with their counterparts in other headquarters. Commanders ensure staff integration through developing the unit's battle rhythm, including synchronizing various meetings, working groups, and boards.

4-28. The basis for staff organization depends on the mission, each staff's broad areas of expertise, and regulations and laws. While staffs at every echelon and type of unit are structured differently, all staffs share some similarities. (See FM 6-0 for an expanded discussion of staffs.)

Liaison Officers

4-29. Liaison is that contact or intercommunication maintained between elements of military forces or other agencies to ensure mutual understanding and unity of purpose and action. Most commonly used for establishing and maintaining close communications, liaison continuously enables direct, physical communications between commands and with unified action partners. Commanders use liaison during operations and normal daily activities to facilitate a shared understanding and purpose among organizations, preserve freedom of action, and maintain flexibility. Liaison provides commanders with relevant information and answers to operational questions, thus enhancing the commander's situational understanding.

4-30. Liaison activities augment the commander's ability to synchronize and converge all elements of combat power into their concept of operation and scheme of maneuver. They include establishing and maintaining physical contact and communications between elements of military forces and nonmilitary agencies during operations. Liaison activities ensure—

- Cooperation and understanding among commanders and staffs of different headquarters.
- Coordination on tactical matters to achieve unity of effort.
- Synchronization of lethal and nonlethal effects.
- Understanding of implied or inferred coordination measures to achieve synchronized results.

4-31. A liaison officer represents a commander or staff officer. Liaison officers transmit information directly, bypassing headquarters and staff layers. A trained, competent, trusted, and informed liaison officer, either a commissioned or a noncommissioned officer, is the key to effective liaison. Liaison officers must have the commander's full confidence and sufficient experience for the mission. At higher echelons, the complexity of operations often requires more senior ranking liaison officers.

4-32. Based on the situation, commanders may receive or request liaison elements (individuals and teams) to assist them with command and control. Liaison elements include but are not limited to—

- An air liaison officer.
- A naval gunfire liaison officer.
- An Army space support team.
- A psychological operations planner.
- An Army cyberspace operations support team.
- An Army field support team.
- A digital liaison detachment
- Liaisons to and from subordinate, adjacent, and supporting units.
- Liaisons to and from unified action partners.

(See FM 6-0 for more discussion on liaison.)

Processes

4-33. Commanders establish and use systematic processes and procedures to organize the activities within their headquarters and throughout the force. Processes are a series of actions directed to an end state, such as the military decision-making process. In addition to the major activities of the operations process, commanders and staffs use several integrating processes to synchronize specific functions throughout the operations process. The integrating processes are—

- Intelligence preparation of the battlefield (described in ATP 2-01.3).
- Information collection (described in FM 3-55).
- Targeting (described in ATP 3-60).
- Risk management (described in ATP 5-19).
- Knowledge management (described in chapter 3)

4-34. Procedures govern actions within the command and control system to make it more effective and efficient. They allow for complex actions to take place without detailed guidance every time the procedure is initiated. For example, standard operating procedures often provide detailed unit instructions on how to configure COP displays. Adhering to processes and procedures minimizes confusion, misunderstanding, and hesitation as commanders make frequent, rapid decisions to meet operational requirements.

4-35. Procedures can increase organizational competence by improving a staff's efficiency or by increasing the tempo. Procedures can be especially useful in improving the coordination of Soldiers who must cooperate to accomplish repetitive tasks, such as the internal functioning of a command post.

4-36. Command and control procedures are designed for simplicity and speed: they should be simple enough to perform quickly and smoothly under conditions of extreme stress. They should be efficient enough to increase tempo. Streamlined staff-planning sequences are preferable to deliberate, elaborate ones.

4-37. Commanders establish procedures to streamline operations and written orders and help integrate new Soldiers and attachments. Usually spelled out in unit standard operating procedures, procedures also help commanders make decisions faster by providing relevant information in standard, easy-to-understand formats. Procedures describe routine actions, thus eliminating repetitive decisions, such as, where to put people in a command post, how to set up a command post, and march formations.

4-38. Procedures facilitate continuity of operations when leaders become unable to perform their duties. Subordinates can step in and use procedures to continue to operate. When Soldiers are tired or stressed, their decision making capability is degraded. Standard operating procedures help individuals and units continue to accomplish many tasks because they are routine.

4-39. Procedures do not cover every possible situation. Units avoid applying procedures blindly to the wrong tasks or the wrong situations, which can lead to ineffective, even counterproductive, performance.

Networks

4-40. Networks in the command and control system collect, process, store, display, disseminate, and protect information worldwide. They enable the execution of command and control, and they support operations through the wide dissemination of data and relevant information.

4-41. Networks enable commanders to communicate information and control forces whether mounted or dismounted, and they are key enablers of successful operations. Commanders systematically establish networks to connect people and allow sharing of information and resources. The Army's primary network is the Department of Defense information network-Army. This network consists of—

- End-user applications.
- Information services and data.
- Network transport and management.

End-User Applications

4-42. Commanders determine their information requirements and focus their staffs and organizations on using the application layer to meet these requirements. End-user applications include automated information systems, software, and user devices that allow users to display and disseminate information, and the policies and procedures for their use. This includes telephones, tablets, laptops, software applications, and user interfaces. End-user applications allow people to leverage the network's capabilities.

4-43. End-user applications directly affect how commanders communicate and how their staffs collaborate. They promote unity of effort by allowing commanders to view and understand their areas of operation, communicate the commander's intent, and disseminate relevant information. When operated by well-trained personnel and used properly, these applications can give commanders an information advantage over opponents by reducing the time required to make a decision, improving combined arms coordination, and synchronizing the warfighting functions. Applications can simultaneously support current and future operations as well as plans.

Information Services and Data

4-44. The primary purpose of information services and data is to facilitate timely and accurate decision making and execution by processing and managing information. The services and data include all the information services, servers, and data standards that collect, process, and store information. This includes the servers, data storage and distribution, cloud and edge servers, and software and data standards that allow the display of a COP as meaningful visual images that directly impart knowledge and increase understanding.

4-45. Information services and data include the information management processes that support knowledge management. Information and knowledge management reduce the time and effort commanders spend assimilating information and developing situational understanding. Information services and data support shared situational understanding.

Network Transport and Management

4-46. Network transport is processes, equipment, and transmission media that provide connectivity and move data between networking devices and facilities. *Network transport* is a system of systems including the people, equipment, and facilities that provide end-to-end communications connectivity for network components (FM 6-02). The primary purpose of network transport is to move data between networking devices and facilities. Network transport devices include radios, Wi-Fi, microwave and satellite communications, and cable and wire.

4-47. Network management equipment controls the movement of data around the battlefield. Network management devices include switches, routers, and communications security equipment. Network management also includes software designed to operate and secure all aspects of the network, including end-user applications, information services, data, and network transport. This software is distinct from end- user applications in that it provides functions to control and secure the network, rather than user services.

4-48. Successful commanders understand that networks may be degraded through threat or environmental factors during operations. They develop methods and measures to mitigate the impact of degraded networks. This mitigation may be through exploiting the potential of technology or through establishing trust, creating shared understanding, or providing a clear intent using mission orders.

4-49. Effective commanders use technology to enable a mission command approach to command and control, not to micromanage operations. Equipment that improves the ability to monitor the situation at lower levels increases the temptation to directly control subordinates' actions and thereby undermine mission command. Moreover, such use tends to fix the higher echelon commander's attention at too low a level. Commanders who focus at too low a level risk losing sight of the larger overall picture. Consequently, increased network capabilities bring the need for increased understanding and discipline. Just because technology allows detailed supervision does not mean commanders should employ it in that manner. Effective mission command requires senior commanders to give the on-scene commander freedom to exercise initiative.

COMMAND POSTS

4-50. Effective command and control requires continuous, and often immediate, close coordination, synchronization and information sharing across the staff. To promote this, commanders organize their staffs and other components of the command and control system into command posts to assist them in effectively conducting operations. A *command post* is a unit headquarters where the commander and staff perform their activities (FM 6-0). Often divided into echelons, each echelon of the headquarters is a command post regardless of whether the commander is present or not. When necessary, commanders control operations from other locations away from the command post. In all cases, the commander alone exercises command when in a command post or elsewhere.

4-51. Command posts are facilities that include personnel, processes and procedures, and networks that assist commanders in command and control. Commanders employ command posts to help control operations through continuity, planning, coordination, and synchronizing of the warfighting functions. Commanders organize their command posts flexibly to meet changing situations and requirements of different operations.

4-52. Command post functions directly relate to assisting commanders in understanding, visualizing, describing, directing, leading, and assessing operations. Different types of command posts, such as the main command post or the tactical command post, have specific functions by design. Functions common to all command posts include—

- Conducting knowledge management, information management, and foreign disclosure.
- Building and maintaining situational understanding.
- Controlling operations.
- Assessing operations.
- Coordinating with internal and external organizations.
- Performing command post administration.

Conducting Knowledge Management, Information Management, and Foreign Disclosure

4-53. When combined, knowledge management, information management, and foreign disclosure enable the provision of relevant information to the right person at the right time and in a usable format. Knowledge management, information management, and foreign disclosure facilitate understanding and decision making.

4-54. Commanders and unified action partners must receive combat information and intelligence products in time and in an appropriate format to facilitate shared understanding and support decision making. Timely dissemination of information is critical to the success of multinational operations. Dissemination is deliberate and ensures consumers receive combat information and intelligence products to support operations. Therefore, information should be shared to the maximum extent allowed by law, regulation, and government-wide policy.

4-55. Developing and managing a unit's battle rhythm is a key aspect of effective knowledge management. A unit's battle rhythm establishes various meetings, working groups, and planning teams to assist commanders and staffs with integrating the warfighting functions, coordinating activities, and making effective decisions throughout the operations process. The battle rhythm arranges the sequence and timing of reports, meetings, and briefings based on the commander's preference, higher headquarters requirements, and the type of operations. There is no standard battle rhythm for all units. Depending on echelon and type of operations, commanders and staffs develop and adjust their battle rhythm based on the situation. Managed by the chief of staff or executive officer, a unit's battle rhythm facilitates decision making and routine interactions among commanders, staffs, forces, and unified action partners. (See ATP 6-0.5 for techniques on building a unit's battle rhythm.)

Building and Maintaining Situational Understanding

4-56. Effective knowledge management and information management are essential to building and maintaining situational understanding. Building and maintaining situational understanding helps in establishing the situation's context, developing effective plans, assessing operations, and making quality decisions during execution. Command post activities that contribute to this include—

- Receiving information, including reports from subordinate units.
- Analyzing information.
- Generating, distributing, and sharing information and knowledge products, to including reports required by higher headquarters.
- Conducting battle tracking.
- Conducting update and information briefings.

4-57. Running estimates and the COP are key products used for building and maintaining situational understanding. A *running estimate* is the continuous assessment of the current situation used to determine if the current operation is proceeding according to the commander's intent and if planned future operations are supportable (ADP 5-0). In their running estimates, each command post cell and staff section continuously considers the effects of new information and they update—

- Facts.
- Assumptions.
- Friendly force status.
- Enemy activities and capabilities.
- Civil considerations.
- Conclusions and recommendations.

4-58. The staff uses its running estimates to advise the commander and make recommendations. Information in running estimates also helps build the COP. Maintaining the COP within a command post and with other command posts assists commanders and staffs in maintaining situational understanding and promoting a shared understanding throughout the command.

Controlling Operations

4-59. Personnel within command posts assist commanders in controlling operations, including coordinating, synchronizing, and integrating actions within their delegated authority. They also integrate and synchronize resources in accordance with their commander's priority of support. Staff members monitor and evaluate the progress of operations and make or recommend adjustments to operations in accordance with their commander's intent. While all command posts assist commanders in controlling operations, different command posts are assigned specific control responsibilities. For example, a brigade commander may employ the brigade tactical command post to control battalion air assault operations.

Assessing Operations

4-60. Personnel within command posts continuously assess operations. *Assessment* is the determination of the progress toward accomplishing a task, creating a condition, or achieving an objective (JP 3-0). Assessment involves deliberately comparing forecasted outcomes with actual events to determine the overall effectiveness of force employment. More specifically, assessment helps commanders and staffs in determining progress toward attaining the desired end state, achieving objectives, and performing tasks. It also involves continuously monitoring and evaluating the operational environment to determine which changes might affect the conduct of operations. (See ADP 5-0 for a detailed discussion of assessment.)

Coordinating with Internal and External Organizations

4-61. Units do not operate in isolation. They synchronize their actions with those of others. Coordination is essential to this synchronization. Personnel within command posts continuously coordinate with higher echelon, lower echelon, adjacent, supporting, and supported units, and with unified action partners. Coordination helps—

- Develop shared understanding.
- Ensure a thorough understanding of the commander's intent and concept of operations.
- Inform an organization on issues so that it may adjust plans and actions as required.
- Avoid conflict and duplication of effort among units.
- Ensure synchronization of effects and efforts between supporting and supported units.

Performing Command Post Administration

4-62. Commanders staff, equip, and organize command posts to support 24-hour operations. As such, command post personnel and equipment must be protected and sustained. This requires an effective standard operating procedure and personnel trained on command post administration, including—

- Establishing the command post.
- Displacing the command post.
- Providing security.
- Maintaining continuity of operations.
- Executing sleep plans.
- Managing stress.

Organization and Employment Considerations

4-63. Command posts provide locations from which commanders, assisted by their staffs, command operations and integrate and synchronize combat power to accomplish missions across the range of military operations. Commanders organize the other components of the command and control system into command posts based on mission requirements and the situation that will best assist them in exercising mission command. Planning considerations for command post organization and employment can be categorized as—

- Those contributing to effectiveness.
- Those contributing to survivability.

In many cases, these factors work against each other and therefore neither can be optimized. Tradeoffs are made to acceptably balance effectiveness and survivability.

Effectiveness

4-64. Command post personnel, equipment, and facilities are arranged to facilitate coordination, exchange of information, and rapid decision making. A command post must effectively communicate with higher echelon, subordinate, adjacent, supporting, and supported units and have the ability to move as required. Considerations for command post effectiveness include design layout, standardization, continuity, and capacity.

Design Layout

4-65. Well-designed command posts integrate command and staff efforts. Within a command post, the location of staffs are arranged to facilitate internal communication and coordination. This arrangement may change over the course of operations as the situation changes. Other layout considerations include—

- The ease of information flow.
- User interface with communications systems.

- The positioning of information displays for ease of use.
- The integrating of complementary information on maps and displays.
- Adequate workspace for the staff and commander.
- The ease of displacement (including setup, tear-down, and movement).

Standardization

4-66. Standardization increases the efficiency of command post operations. Commanders develop detailed standard operating procedures for all aspects of command post operations, including command post layout, battle drills, meeting requirements, and reporting procedures. Command post standard operating procedures are enforced and revised throughout training. Doing this makes many command post activities routine. Trained staffs are prepared to effectively execute drills and procedures in demanding, stressful times during operations.

Continuity

4-67. Command posts must be manned, equipped, and organized to control operations without interruptions. Commanders organize command posts in order to support continuous operations. To support continuous operations, unit standard operating procedures address shift plans, rest plans, and procedures for loss of communications with the unit commander, subordinates, or other command posts.

4-68. Maintaining continuity during displacement of a command post or catastrophic loss requires designating alternate command posts and passing control between command posts. Continuity of command requires commanders to designate seconds in command and inform them of all critical decisions. Primary staff officers should also designate alternates.

Capacity

4-69. Command posts should be manned and organized to manage the information needed to operate effectively. The capacity to plan, prepare, execute, and continuously assess operations concerns both staffing and the network, as does the ability to manage relevant information. Command post personnel must be trained and have the requisite tactical and technical proficiency.

Survivability

4-70. Command post survivability is vital to mission success and is measured by the capabilities of the threat in the context of the situation. Survivability may be obtained at the price of effectiveness. Depending on the threat, command posts need to remain small and highly mobile—especially at lower echelons. Command posts are easily acquired and targeted when they are concentrated. Considerations for command post survivability include dispersion, size, redundancy, mobility, electronic and thermal signatures, and camouflage and concealment. Additional measures include cover or shielding by terrain features or urban structures. (See ATP 3-37.34 for more information on command post survivability.)

Dispersion

4-71. Dispersing command posts enhances the survivability of the commander's command and control system. Commanders place minimum resources forward and keep more elaborate facilities back. This makes it harder for enemies to find and attack them. It also decreases support and security requirements forward. Depending on the situation, commanders may leave personnel and equipment at home station to perform detailed analysis and long-range planning for operations.

Size

4-72. A command post's size affects its mobility and survivability. Large command posts can increase capacity and ease face-to-face coordination. Their size, however, makes them vulnerable to multiple types of acquisitions and attack. Smaller command posts are easier to protect, but they may lack capacity to control operations effectively. The key to success is achieving the right balance.

Redundancy

4-73. Reducing command post size cuts signature and enhances mobility. However, some personnel and equipment redundancy is required for continuous operations. In operations, personnel and equipment are lost or fail under stress. Having the right amount of redundancy allows command posts to continue to operate effectively when this happens.

Mobility

4-74. Command posts must deploy efficiently and move within the area of operations as the situation requires. Command post mobility is important, especially at lower echelons during combat operations. Lower-echelon command posts and those employed forward in the combat zone may need to move quickly and often. Both small size and careful transportation planning facilitates rapid displacement of command posts.

ORGANIZING FOR COMMAND AND CONTROL

4-75. How commanders organize a command and control system can complicate or simplify execution. Organizational decisions establish the chain of command (command and support relationships) and task organization. They can influence where commanders obtain information, whom they rely on for advice, and how they supervise execution of their decisions. Organizational decisions affect the structure of the flow of recommendations to commanders. In large part, the organization establishes formal communication channels and determines how commanders distribute information throughout their forces.

4-76. Commands operate most efficiently and effectively when Soldiers consider themselves part of a team or larger organization. Organization serves the important function of providing sources of group identity for soldiers assigned to a command. A command operates most effectively when soldiers consider themselves members of one or more groups characterized by high levels of loyalty, cooperation, morale, and commitment. This supports mission command.

4-77. Information flows vertically within the chain of command, but the organization should not limit its flow to the chain of command. Information also must flow laterally among adjacent, supported, and supporting units. Information flows informally and unofficially—between individuals according to personal relationships—as well as within formal channels. Informal channels provide important redundancy.

GUIDES TO ORGANIZING FOR COMMAND AND CONTROL

4-78. When organizing for command and control, commanders consider—

- The chain of command.
- Span of control.
- Unit integrity.
- Degraded environments.

The Chain of Command

4-79. The *chain of command* is the succession of commanding officers from a superior to a subordinate through which command is exercised (JP 1). The commander at each level responds to orders from a higher commander and, in turn, issues orders to subordinates. In this way, the chain of command fixes responsibility and sources of authority at each echelon while, at the same time, distributing them broadly throughout the force. Each commander has designated authority and responsibility in a given area of operations or area of responsibility. Command and support relationships specify the type and degree of authority one commander has over another, and the type and degree of support one commander provides to another.

Command and Support Relationships

4-80. Establishing clear command and support relationships is fundamental to organizing for operations. These relationships prescribe clear responsibilities and authorities among subordinate and supporting units. Some forces are given command or support relationships that limit their commander's authority to prescribe additional relationships. Knowing the inherent responsibilities of each command and support relationship allows commanders to establish clear responsibilities when organizing their forces. (See ADP 3-0 for more information on command and support relationships.)

4-81. Commanders designate command and support relationships within their authority to weight the decisive operation and support the concept of operations. Task organization also helps subordinate and supporting commanders understand their roles in the operation and contribute to achieving the commander's intent. Command and support relationships carry with them varying responsibilities to subordinate units by parent and gaining units.

Allocating Resources

4-82. Mission command requires commanders to have authority over or access to all resources required to accomplish the mission. Accordingly, commanders organize resources as well as forces when making organizational decisions. This resource organization may be implicit in the command and support relationships established; however, it may differ partly or completely from them, as in establishing priorities including fires, work, or sustainment. In any case, the resource organization must not violate unity of command and should support unity of effort. Further, this organization or allocation of resources should have minimum restrictions on their use, permitting subordinates to further reallocate or to employ them as the tactical situation requires.

Span of Control

The average human brain finds its effective scope in handling three to six other brains.

General Sir Ian Hamilton

4-83. Organization should ensure reasonable span of control, which refers to the number of subordinates or activities under the control of a single commander. A commander's span of control should not exceed that commander's capability to command effectively. The optimal number of subordinates is situation-dependent. The more fluid and fast-changing the situation, the fewer subordinate elements a commander can supervise closely. Within this situation-dependent range, a greater number of subordinates allows greater flexibility, and increases options and combinations. However, as the number increases, commanders, at some point, lose the ability to consider each unit individually and begin to think of the units as a single, inflexible mass. At this point, the only way to reintroduce flexibility is to group elements into a smaller number of parts, creating another echelon of command.

4-84. Narrowing the span of control—that is, lessening the number of immediate subordinates—deepens the organization by adding layers of command. The more layers of command in an organization, the longer it takes for information to move up or down the organization. Consequently, the organization may become slower and less responsive. Conversely, an effort to increase tempo by eliminating echelons of command or flattening an organization necessitates widening the span of control. Commanders balance width and depth, so that the command and control organization fits the situation.

4-85. An effective organization enables the commander and subordinate commanders to command without information overload. The commander establishes his span of control and organizes the command and control system so as to be able to exercise command and control under all circumstances, including degraded environments.

Unit Integrity

4-86. Mission command requires subordinate commands capable of operating in the absence of detailed orders. Forming such task-organizations increases each commander's freedom of action. Effective commanders are flexible: they task-organize forces to suit the situation. This might include creating nonstandard, temporary teams or task forces. However, commanders reconcile the need for organizational flexibility with the requirement to create shared understanding and mutual trust. These characteristics result from familiarity and stable working relationships.

4-87. One way to balance these demands is to observe unit integrity when organizing for command and control. Commanders must take into account the impact on mission command when task-organizing forces.

Whenever possible, commanders should task-organize based on standing headquarters and habitually associated elements. When this is not feasible and organizations are formed from a wide variety of units, commanders must allow time for training and establishing functional working relationships and procedures.

4-88. Once a force is task-organized and committed, commanders do not change the task-organization during operations unless the benefits clearly outweigh the disadvantages. Reorganizations cost time, momentum, effort, and tempo. Commanders also consider logistical factors, as the time required to change task-organization may counter any organizational advantages.

Degraded Environments

4-89. When organizing for command and control, commanders consider the impact of degraded environments on the command and control system. The command and control system may be degraded as the result of hostile actions to contest the use of space and the information environment or due to the lack of resources to provide sufficient network coverage in an area of operations. The degradation may not be technological. The use of chemical, biological, radiological, nuclear, and high-yield explosives, or adverse weather, may create physical conditions that cause interference in the electromagnetic spectrum and degrade the command and control system. All of these may interfere with a commander's ability to exercise command and control.

4-90. In order to mitigate this risk and successfully conduct operations in degraded environments, commanders cannot become over reliant on technological capabilities. Commanders ensure their personnel are trained on analog and manual command and control processes and are comfortable operating in degraded environments. Understanding the functions performed by automated systems is critical to understanding what functions must be performed in an analog environment. Commanders ensure standard operating procedures are in place that will help the command and control system maintain its ability to function. Personnel should be trained and proficient in—

- Continuous operations.
- Maintaining the COP.
- Manual information sharing.
- Staff integration and crosstalk.
- Manual running estimates.
- Command post battle drills.

4-91. One way to deal with degraded communications is through primary, alternate, contingency, and emergency (PACE) communication planning. A PACE plan establishes the primary, alternate, contingency, and emergency methods of communications, typically from higher echelons to lower echelons. Establishing a PACE plan requires care that an alternate or contingency method of communications does not rely on the primary. The key to a good PACE plan is to establish redundancy so that communications are always available. Most units will have two PACE plans: one for communications to higher echelon headquarters and one for subordinate units. A PACE plan for a higher echelon headquarters will likely be established by that headquarters. (See ATP 6-0.5 for more information on PACE planning.)

4-92. Ultimately, the doctrinal solution to operating in degraded environments is mission command. Even under severely degraded conditions, Army forces continue to make decisions and act in the absence of orders, when existing orders no longer fit the situation, or when unforeseen opportunities arise.

CONCLUSION

4-93. Commanders alone cannot exercise command and control. At each echelon of command, commanders have a command and control system to provide support. That system is more than equipment; it consists of all the resources available to commanders to help them exercise authority and direction. How commanders organize, locate, and design their command and control systems directly affect their ability to conduct operations. A properly designated commander and a well-designed command and control system provides for continuity of command and control.

This page intentionally left blank.

Source Notes

This division lists sources by page number. Where material appears in a paragraph, it lists both the page number followed by the paragraph number.

viii "An order should…": *Field Service Regulation* (obsolete) (Washington DC: Government Printing Office, 1905), 29.

viii "These thunder runs…": Lieutenant General David G. Perkins, "Mission Command: Reflections from the Combined Arms Commander," *Army Magazine*, Volume 62 (June 2012), 32.

1-1 "The situations that…": FM 100-5, *Field Service Regulations: Operations* (obsolete) (Washington DC: Government Printing Office, 1941), 24.

1-2 "War is the …": Carl von Clausewitz, *On War*, trans. and ed. M. Howard and P. Paret (Boston, MA: Princeton University Press, 2004), 101–102.

1-2 "The role of …": B.H. Liddell Hart, *Strategy*, 2d rev. ed. (Toronto, Canada: Meridian, 1991), 321-22.

1-3 "Never tell people …": General George S. Patton, *War as I Knew It* (Boston, MA: Houghton Mifflin Company, 1947), 357.

1-4 **Von Moltke and Auftragstaktik** Adapted from Helmuth von Moltke, *Moltke's Military Works*, Vol. 4, War Lessons, Part I, "Operative Preparations for Battle," trans. Harry Bell (Fort Leavenworth, KS: Army Service Schools, 1916), 65–67.

1-9 **Command Based on Shared Understanding and Trust: Grant's Orders to Sherman, 1864.** Ulysses S. Grant, *Personal Memoirs of U. S. Grant* (New York, NY: Charles L. Webster & Company, 1885–1886). Online https://www.fulltextarchive.com/page/Personal-Memoirs-of-U-S-Grant--Volume-Two9/. *War of the Rebellion: Serial 059* Pages 0313–0315 Chapter XLIV. CORRESPONDENCE, ETC. - UNION. (no date). Retrieved from https://ehistory.osu.edu/books/official-records/059/0313.

1-9 "I suppose dozens of …": William Slim, *Defeat Into Victory: Battling Japan in Burma and India, 1942–1945* (London: Casell, 1956; reprint New York, NY: Cooper Square Press, 2000), 210–211.

1-10 "An order should not …": FM 100-5. *Tentative Field Service Regulations: Operations* (obsolete) (Washington DC: Government Printing Office, 1939), 109.

1-11 "Every individual from the …": FM 100-5. *Tentative Field Service Regulations: Operations* (obsolete) (Washington, DC: Government Printing Office, 1941), 32.

1-13 **Initiative: U.S. Paratroopers in Sicily.** Logan Nye, "How the 'Little Groups of Paratroopers' Became Airborne Legends," *We Are the Mighty*. Posted on 8 April 2016. Online http://freerepublic.com/focus/f-chat/3535576/posts?page=12.

1-13 "Given the same …": Carl Von Clausewitz, *On War*, trans. and ed. M. Howard and P. Paret (Boston, MA: Princeton University Press, 2004), 191.

1-15 **Corporal Alvin York and Mission Command.** Vignette adapted from Douglas V. Mastriano, "Thunder in the Argonne! SGT Alvin York and Mission Command," *INFANTRY* (July–September 2015), 71–75.

1-16 "If intercommunication between …": J. F. C. Fuller, *Infantry in Battle* (Washington, DC: The Infantry Journal, Incorporated, 1939), 179.

1-16 "I believe firmly in …": Field-Marshal Earl Wavell, *Soldiers and Soldiering or Epithets of War* (Oxford, United Kingdom: Alden Press, 1953), 127.

2-1 "When you are…": George C. Marshall: in *Selected Speeches and Statements of General of the Army George C. Marshall*, ed. H.A. DeWeerd (Washington, D.C.: *The Infantry Journal*, 1945), 176.

2-2 **Assuming Command: General Ridgway Takes Eighth Army.** Matthew B. Ridgway, *Soldier: The Memoirs of Matthew B. Ridgway* (New York, NY: Harper, 1956; reprinted by Andesite Press, 2017), 195-199.

2-3 "The commander must …": *Truppen Fuhrung: German Field Regulations, Volume 1*, (1935), paragraph 37.

2-5 "Intuition depends on…": Gary Klein, *Sources of Power: How People Make Decisions* (Cambridge, MA: The MIT Press, 1999), 33.

2-6 "Despite the years…": General George S. Patton, "The Secret of Victory," speech given on 26 March 1926. *Military Essays and Articles by George S. Patton, Jr. General, U.S. Army 02605 1885 – 1945*. ed. Charles M. Province. San Diego, California: The George S. Patton, Jr. Historical Society, 2002. Online http://www.pattonhq.com/pdffiles/vintagetext.pdf.

2-8 **Risk Acceptance: OPERATION HAWTHORN, Dak To, Vietnam**. Based on John M. Carland, *Combat Operations: Stemming the Tide*, May 1965 to October 1966, The United States Army in Vietnam, CMH Pub 91-5 (Washington, DC: Center of Military History, US Army: GPO, 2000), 277–287.

2-8 "If it is necessary …": Field Marshall Carver cited in ADP AC 71940, *Land Operations* (Bristol, United Kingdom: Land Warfare Development Centre, 2017), 9-7.

2-10 "As each man's…": Carl Von Clausewitz, *On War*, trans. and ed. M. Howard and P. Paret (Boston, MA: Princeton University Press, 2004), 104.

2-11 "One of the…": William Joseph Slim, *Unofficial History* (New York, NY: Orion Publishing Group, 1962), 156.

2-12 "Commanders must remember…": General George S. Patton, *War as I Knew It* (Boston, MA: Houghton Mifflin Company, 1947), 357.

2-16 "There will be…": Admiral E. J. King, CINCLANT Serial 053, 21 January 1941. Online https://www.usnwc.edu/Portals/16/PCO%20Alumni%20Content/ADM-King-Serial-053.pdf?ver=2017-10-23-121408-920.

2-16 "Morale is a state…": William Joseph Slim, *Defeat Into Victory: Battling Japan in Burma and India, 1942–1945* (London: Casell, 1956; reprint New York, NY: Cooper Square Press, 2000), 182.

2-17 "Judgment comes from experience…": Simon Bolivar Buckner, as quoted by Omar N. Bradley, "Leadership: An Address to the US Army War College, 07 Oct. 71," *Parameters 1* (3) (1972): 8.

2-18 **Mutual Trust and Shared Understanding: VII Corps and the Ruhr Encirclement.** William M. Connor "Establishing Command Intent, a Case Study: The Encirclement of the Ruhr, March 1945" in *The Human in Command: Exploring The Modern Military Experience*, ed. Carol McCann and Ross Pigeau. Toronto, Canada: Kluwer Academic/Plenum Press, 2000.

2-20 "General Meade was…": Ulysses S. Grant, *Memoirs and Selected Letters: Personal Memoirs of U.S. Grant, Selected Letters, 1839-1865*, vol. 2, ed. William S. McFeely and Mary Drake McFeely (New York, NY: Library of America, 1990), 770.

2-23 "No man is …": *Genghis Khan: The Emperor of All Men*, ed. Harold Lamb (New York, NY: Robert McBride, 1927; reprint, New York: Doubleday, 1956), 46.

2-23 "I have found again …": Erwin Rommel. *The Rommel Papers,* ed. B. H. Liddell Hart (New York, NY: Harcourt, Brace, 1953), 7.

2-24 "Good morale and…": FM 100-5, *Operations* (obsolete) (Washington, DC: Army Publishing Directorate, 1941), 20–21.

3-1 "The test of …": J.F.C. Fuller, as quoted in *Infantry in Battle* (Washington, DC: The Infantry Journal, Incorporated, 1939; reprint, Fort Leavenworth, KS: US Army Command and General Staff College: GPO, 1981), 169.

3-1 "Everything in war ....": Carl von Clausewitz, *On War*, trans. and ed. M. Howard and P. Paret (Boston, MA: Princeton University Press, 2004), 119 and 121.

3-2 **Levels of Control and German Auftragstaktik.** Adapted from Robert A. Doughty, *The Breaking Point: Sedan and the Fall of France, 1940* (Hamden, CT: The Shoe String Press, Inc., 1990), 32–36.

3-5 "Many intelligence reports in....": Carl von Clausewitz, *On War*, trans. and ed. M. Howard and P. Paret (Boston, MA: Princeton University Press, 2004), 117.

3-6 "It is in ...." B. H. Liddell Hart cited in *Dictionary of Military and Naval Quotations*, ed. Robert D. Heinl Jr., (Annapolis, MD: US Naval Institute, 1966), 61.

3-7 "He who wars walks ...": Sir William Napier cited in *Dictionary of Military and Naval Quotations*, ed. Robert D. Heinl Jr., (Annapolis, MD: US Naval Institute, 1966), 61.

3-10 **Crosstalk in the Desert-VII Corps in the Gulf War.** Based on TRADOC Pam 525-100-1, *Leadership and Command on the Battlefield: Operations JUST CAUSE and DESERT STORM* (obsolete) (Fort Monroe, VA: HQ, TRADOC, 1992), 28.

3-14 "...avoid taking...." paraphrased from Richard E. Simpkin and John Erickson, *Deep Battle: The Brainchild of Marshal Tukhachevskii* (London: Brassey's Defence, 1987), 150.

3-16 "A doctrine of war ...": Ferdinand Foch cited in *Dictionary of Military and Naval Quotations*, ed. Robert D. Heinl Jr., (Annapolis, MD: US Naval Institute, 1966), 95.

4-1 "Staff systems and...": General George S. Patton, Jr. "The Secret of Victory." 1926 cited in *Military Essays and Articles by George S. Patton, Jr. General, U.S. Army 02605 1885 – 1945*. ed. Charles M. Province. San Diego, California: The George S. Patton, Jr. Historical Society, 2002. Online http://www.pattonhq.com/pdffiles/vintagetext.pdf.

4-3 "When placed in...": General H. Norman Schwarzkopf cited in Johnson, Erin. "Schwarzkopf speaks of leadership at symposium." *The Daily Universe*. 21 October 2001. Online https://universe.byu.edu/2001/10/11/schwarzkopf-speaks-of leader at symposium/.

4-3 "A lazy commander...": Address by Lt.-Col. Simonds, Commandant, Canadian Junior War Staff Course, 12 April 1941 (as related by Major C.P Stacey, Historical Officer, C.M.H.Q.). Online https://www.canada.ca/en/department-national-defence/services/military-history/history-heritage/official-military-history-lineages/reports/military-headquarters-1940-1948/closing-exercises-canadian-junior-war-staff-course.html.

4-14 "The average human...": General Sir Ian Hamilton, *The Soul and Body of an Army* (London, United Kingdom: Edward Arnold & Co., 1921), 229.

Glossary

The glossary lists acronyms and terms with Army or joint definitions. Where Army and joint definitions differ, (Army) precedes the definition. Terms for which ADP 6-0 is the proponent are marked with an asterisk (*). The proponent publication for other terms is listed in parentheses after the definition.

SECTION I – ACRONYMS AND ABBREVIATIONS

1SG	first sergeant
ADP	Army doctrine publication
AR	Army regulation
ATP	Army techniques publication
C2	command and control
CCIR	commander's critical information requirement
CJCSM	Chairman of the Joint Chiefs of Staff manual
COP	common operational picture
CPL	corporal
CPT	captain
DA	Department of the Army
FM	field manual
GA	General of the Army
GEN	general
JP	joint publication
LTG	lieutenant general
MG	major general
NVA	North Vietnamese Army
PACE	primary, alternate, contingency, and emergency
SGT	sergeant
U.S.	United States

SECTION II – TERMS

Army team building

A continuous process of enabling a group of people to reach their goals and improve effectiveness through leadership and various exercises, activities and techniques. (FM 6-22)

assessment

The determination of the progress toward accomplishing a task, creating a condition, or achieving an objective. (JP 3-0)

chain of command

The succession of commanding officers from a superior to a subordinate through which command is exercised. (JP 1)

***civil considerations**

The influence of manmade infrastructure, civilian institutions, and attitudes and activities of the civilian leaders, populations, and organizations within an area of operations on the conduct of military operations.

combat power

(Army) The total means of destructive, constructive, and information capabilities that a military unit or formation can apply at one time. (ADP 3-0)

command

The authority that a commander in the armed forces lawfully exercises over subordinates by virtue of rank or assignment. (JP 1)

command and control

The exercise of authority and direction by a properly designated commander over assigned and attached forces in the accomplishment of the mission. (JP 1)

***command and control system**

(Army) The arrangement of people, processes, networks, and command posts that enable commanders to conduct operations.

command and control warfighting function

The related tasks and a system that enable commanders to synchronize and converge all elements of combat power. (ADP 3-0)

commander's critical information requirement

An information requirement identified by the commander as being critical to facilitating timely decision making. (JP 3-0)

commander's intent

A clear and concise expression of the purpose of the operation and the desired military end state that supports mission command, provides focus to the staff, and helps subordinate and supporting commanders act to achieve the commander's desired results without further orders, even when the operation does not unfold as planned. (JP 3-0)

***commander's visualization**

The mental process of developing situational understanding, determining a desired end state, and envisioning an operational approach by which the force will achieve that end state.

command post

A unit headquarters where the commander and staff perform their activities. (FM 6-0)

***common operational picture**

(Army) A display of relevant information within a commander's area of interest tailored to the user's requirements and based on common data and information shared by more than one command.

***control**

The regulation of forces and warfighting functions to accomplish the mission in accordance with the commander's intent.

***control measure**

A means of regulating forces or warfighting functions.

***data**

In the context of decision making, unprocessed observations detected by a collector of any kind (human, mechanical, or electronic).

***essential element of friendly information**

A critical aspect of a friendly operation that, if known by a threat would subsequently compromise, lead to failure, or limit success of the operation and therefore should be protected from enemy detection.

friendly force information requirement

Information the commander and staff need to understand the status of friendly force and supporting capabilities. (JP 3-0)

***graphic control measure**

A symbol used on maps and displays to regulate forces and warfighting functions.

***information**

In the context of decision making, data that has been organized and processed in order to provide context for further analysis.

***information management**

(Army) The science of using procedures and information systems to collect, process, store, display, disseminate, and protect data, information, and knowledge products.

***key tasks**

Those significant activities the force must perform as a whole to achieve the desired end state.

***knowledge**

In the context of decision making, information that has been analyzed and evaluated for operational implications.

***knowledge management**

The process of enabling knowledge flow to enhance shared understanding, learning, and decision making.

leadership

The activity of influencing people by providing purpose, direction, and motivation to accomplish the mission and improve the organization. (ADP 6-22)

***mission command**

(Army) The Army's approach to command and control that empowers subordinate decision making and decentralized execution appropriate to the situation.

***mission orders**

Directives that emphasize to subordinates the results to be attained, not how they are to achieve them.

multinational operations

A collective term to describe military actions conducted by forces of two or more nations, usually undertaken within the structure of the coalition or alliance. (JP 3-16)

network transport

A system of systems including the people, equipment, and facilities that provide end-to-end communications connectivity for network components. (FM 6-02)

operational approach

A broad description of the mission, operational concepts, tasks, and actions required to accomplish the mission. (JP 5-0)

operational environment

A composite of the conditions, circumstances, and influences that affect the employment of capabilities and bear on the decisions of the commander. (JP 3-0)

operational initiative

The setting of tempo and terms of action throughout an operation. (ADP 3-0)

operations process

The major command and control activities performed during operations: planning, preparing, executing, and continuously assessing the operation. (ADP 5-0)

priority intelligence requirement

An intelligence requirement that the commander and staff need to understand the threat and other aspects of the operational environment. (JP 2-01)

procedures

Standard, detailed steps that prescribe how to perform specific tasks. (CJCSM 5120.01)

***relevant information**

All information of importance to the commander and staff in the exercise of command and control.

running estimate

The continuous assessment of the current situation used to determine if the current operation is proceeding according to the commander's intent and if planned future operations are supportable. (ADP 5-0)

***situational understanding**

The product of applying analysis and judgment to relevant information to determine the relationships among the operational and mission variables.

***understanding**

In the context of decision making, knowledge that has been synthesized and had judgment applied to comprehend the situation's inner relationships, enable decision making, and drive action.

unified action partners

Those military forces, governmental and nongovernmental organizations, and elements of the private sector with whom Army forces plan, coordinate, synchronize, and integrate during the conduct of operations. (ADP 3-0)

unified land operations

The simultaneous execution of offense, defense, stability, and defense support of civil authorities across multiple domains to shape operational environments, prevent conflict, prevail in large-scale ground combat, and consolidate gains as part of unified action. (ADP 3-0)

unity of effort

Coordination and cooperation toward common objectives, even if the participants are not necessarily part of the same command or organization, which is the product of successful unified action. (JP 1)

warfighting function

A group of tasks and systems united by a common purpose that commanders use to accomplish missions and training objectives. (ADP 3-0)

References

All websites accessed on 17 July 2019.

REQUIRED PUBLICATIONS

These documents must be available to intended users of this publication.

DOD Dictionary of Military and Associated Terms. June 2019.

ADP 1-02. *Terms and Military Symbols*. 14 August 2018.

RELATED PUBLICATIONS

These publications are referenced in this publication.

JOINT AND DEPARTMENT OF DEFENSE PUBLICATIONS

Most joint publications are available online: http://www.jcs.mil/doctrine/.

Most DOD publications are available at the Department of Defense Issuances Web site: https://www.esd.whs.mil/DD/.

CJCSM 5120.01A. *Joint Doctrine Development Process*. 29 December 2014.

JP 1. *Doctrine for the Armed Forces of the United States*. 25 March 2013.

JP 2-01. *Joint and National Intelligence Support to Military Operations*. 5 July 2017.

JP 3-0. *Joint Operations*. 17 January 2017.

JP 3-08. *Interorganizational Cooperation*. 12 October 2016.

JP 3-16. *Multinational Operations*. 1 March 2019.

JP 5-0. *Joint Planning*. 16 June 2017.

ARMY PUBLICATIONS

Most Army doctrinal publications are available online: https://armypubs.army.mil/.

ADP 1-1. *Doctrine Primer*. 31 July 2019.

ADP 3-0. *Operations*. 31 July 2019.

ADP 5-0. *The Operations Process*. 31 July 2019.

ADP 6-22. *Army Leadership*. 31 July 2019.

ADP 7-0. *Training*. 31 July 2019.

AR 380-10. *Foreign Disclosure and Contacts with Foreign Representatives*. 14 July 2015.

AR 600-20. *Army Command Policy*. 6 November 2014.

ATP 2-01.3/MCRP 2-3A. *Intelligence Preparation of the Battlefield/Battlespace*. 1 March 2019.

ATP 3-37.34/MCTP 3-34C. *Survivability Operations*. 16 April 2018.

ATP 3-60. *Targeting*. 7 May 2015.

ATP 5-19. *Risk Management*. 14 April 2014.

ATP 6-0.5. *Command Post Organization and Operations*. 1 March 2017.

ATP 6-01.1. *Techniques for Effective Knowledge Management*. 6 March 2015.

ATP 6-22.5. *A Leader's Guide to Soldier Health and Fitness*. 10 February 2016.

ATP 6-22.6. *Army Team Building*. 30 October 2015.

FM 3-0. *Operations*. 6 October 2017.

FM 3-16. *The Army in Multinational Operations*. 8 April 2014.

FM 3-52. *Airspace Control*. 20 October 2016.

FM 3-55. *Information Collection*. 3 May 2013.

FM 6-0. *Commander and Staff Organization and Operations*. 5 May 2014.

FM 6-02. *Signal Support to Operations*. 22 January 2014.

FM 6-22. *Leader Development*. 30 June 2015.

FM 27-10. *The Law of Land Warfare*. 18 July 1956.

Obsolete Publications

This section contains references to obsolete historical doctrine. The Archival and Special Collections in the Combined Arms Research Library (CARL) on Fort Leavenworth in Kansas contains copies. These publications are obsolete doctrine publications referenced for citations only.

Field Service Regulation (obsolete). Washington DC: Government Printing Office, 1905.

FM 100-5. *Tentative Field Service Regulations, Operations.* (obsolete) Washington DC: Government Printing Office, 1939.

FM 100-5. *Field Service Regulations: Operations* (obsolete).Washington DC: Government Printing Office, 1941.

TRADOC Pam 525-100-1, *Leadership and Command on the Battlefield: Operations JUST CAUSE and DESERT STORM* (obsolete) Fort Monroe, VA: HQ, TRADOC, 1992.

Other Publications

Allied Tactical Publication 3.2.2. *Command and Control of Allied Land Forces*. 15 December 2016.

Bradley, General of the Army Omar N. "Leadership: An Address to the US Army War College, 7 October, 1971." *Parameters 1* (3) (1972): 8.

Carland, John M. *Combat Operations: Stemming the Tide*, May 1965 to October 1966. Center of Military History Publication 91-5, *The United States Army in Vietnam*. Washington, DC: Center of Military History, U.S. Army: GPO, 2000.

Carver, Field Marshal Richard. ADP AC 71940, *Land Operations*. Bristol, United Kingdom: Land Warfare Development Centre, 2017.

Clausewitz, Carl von. *On War*, trans. and ed. M. Howard and P. Paret. Boston, MA: Princeton University Press, 2004.

Connor, William M. "Establishing Command Intent, a Case Study: The Encirclement of the Ruhr, March 1945" in *The Human in Command: Exploring The Modern Military Experience*, ed. Carol McCann and Ross Pigeau. Toronto, Canada: Kluwer Academic/Plenum Press, 2000.

Doughty, Robert A. *The Breaking Point, Sedan and the Fall of France, 1940*. Hamden, CT: The Shoe String Press INC, 1990.

Fuller, J.F. C. *Infantry in Battle*. Washington, DC: The Infantry Journal, Incorporated, 1939. Reprint, Fort Leavenworth, KS: US Army Command and General Staff College: GPO, 1981.

Grant Ulysses S. *Ulysses S. Grant, Memoirs and Selected Letters: Personal Memoirs of U.S. Grant, Selected Letters, 1839-1865, vol. 2, ed.* William S. McFeely and Mary Drake McFeely. New York, NY: Library of America, 1990.

Hamilton, General Sir Ian. *The Soul and Body of an Army.* London, United Kingdom: Edward Arnold & CO. 1921.

Heinl, Colonel Robert Debs Jr. *Dictionary of Military and Naval Quotations*. Annapolis, MD: U.S. Naval Institute, 1966.

King, Fleet Admiral Ernest J. CINCLANT Serial 053, 21 January 1941. https://www.usnwc.edu/Portals/16/PCO%20Alumni%20Content/ADM-King-Serial-053.pdf?ver=2017-10-23-121408-920.

Klein, Gary. *Sources of Power: How People Make Decisions*. Cambridge, MA: The MIT Press, 1999.

Lamb, Harold. ed. Genghis Khan: *The Emperor of All Men.* New York: Robert McBride, 1927; reprint, New York: Doubleday, 1956.

Liddell Hart, B.H. *Strategy*, 2d rev. ed. Toronto, Canada: Meridian, 1991.

Marshall, George C. *Selected Speeches and Statements of General of the Army George C. Marshall*, ed. H.A. DeWeerd. Washington, D.C.: The Infantry Journal, 1945.

Mastriano, Douglas V. "Thunder in the Argonne! SGT Alvin York and Mission Command," *Infantry.* Fort Benning, GA: U.S. Army Infantry School. July–September 2015.

Moltke, Helmuth von, *Moltke's Military Works*, Vol. 4, War Lessons, Part I, "Operative Preparations for Battle," trans. Harry Bell, Fort Leavenworth, KS: Army Service Schools, 1916.

Nye, Logan. "How the 'Little Groups of Paratroopers' Became Airborne Legends," *We Are the Mighty*, 8 April 2016. http://freerepublic.com/focus/f-chat/3535576/posts?page=12.

Patton, General George S. *Military Essays and Articles by George S. Patton, Jr. General, U.S. Army 02605 1885 – 1945.* ed. Charles M. Province. San Diego, California: The George S. Patton, Jr. Historical Society, 2002. http://www.pattonhq.com/pdffiles/vintagetext.pdf.

Patton, General George S. *War as I Knew It.* Boston, MA: Houghton Mifflin Company, 1947.

Perkins, Lieutenant General David G. "Mission Command: Reflections from the Combined Arms Commander," *Army Magazine*, Volume 62. June 2012. 32

Ridgway, Matthew B. *Soldier: Memoirs of Matthew B Ridgway.* New York, NY: Harper, 1956; reprinted by Andesite Press, 2017.

Rommel, Field Marshal Erwin. *The Rommel Papers.* ed. B.H. Liddell Hart. New York, NY: Harcourt, Brace, 1953.

Schwarzkopf, General Norman H. quoted by Johnson, Erin. "Schwarzkopf Speaks of Leadership at Symposium." *The Daily Universe.* 21 October 2001. https://universe.byu.edu/2001/10/11/schwarzkopf-speaks-of-leadership-at-symposium.

Simonds, Commandant, *Closing Exercises, Canadian Junior War Staff Course*, CHHQ Report # 22. London, England. 24 April 1941. https://www.canada.ca/en/department-national-defence/services/military-history/history-heritage/official-military-history-lineages/reports.html.

Simpkin, Richard E. and John Erickson. *Deep Battle: The Brainchild of Marshal Tukhachevskii.* London: Brassey's Defence, 1987, 150.

Slim, Field-Marshall Viscount William Joseph. *Defeat into Victory: Battling Japan in Burma and India, 1942–1945.* London: Casell, 1956; reprint New York, NY: Cooper Square Press, 2000.

Slim, Field-Marshall Viscount William Joseph. *Unofficial History.* New York, NY: Orion Publishing Group, 1962.

Truppen Fuhrung: German Field Regulations, Volume 1. Fort Leavenworth, Kansas: Command and General Staff School Press. 1935. http://cgsc.contentdm.oclc.org/cdm/compoundobject/collection/p4013coll7/id/131.

Wavell Field-Marshal Earl, *Soldiers and Soldiering or Epithets of War.* Oxford, United Kingdom: Alden Press, 1953.

PRESCRIBED FORMS

This section contains no entries.

REFERENCED FORMS

Unless otherwise indicated, DA forms are available on the Army Publishing Directorate Website: https://armypubs.army.mil/.

DA Form 2028. *Recommended Changes to Publications and Blank Forms.*

This page intentionally left blank.

Index

Entries are by paragraph number.

Entries are by paragraph number.

Entries are by paragraph number.

Made in the USA
Columbia, SC
29 July 2024